Laboratory Manual In Conceptual Physics

Second Edition
LABORATORY MANUAL IN CONCEPTUAL PHYSICS

BILL W. TILLERY
Arizona State University

WCB Wm. C. Brown Publishers

Dubuque, IA Bogota Boston Buenos Aires Caracas Chicago
Guilford, CT London Madrid Mexico City Sydney Toronto

Book Team

Editor *Jeffrey L. Hahn*
Developmental Editor *Rachel Riley*
Production Editor *Cathy Ford Smith*
Design Freelance Specialist *Barb Hodgson*

Wm. C. Brown Publishers
A Division of Wm. C. Brown Communications, Inc.

Vice President and General Manager *Beverly Kolz*
Vice President, Publisher *Earl McPeek*
Vice President, Director of Sales and Marketing *Virginia S. Moffat*
Vice President, Director of Production *Colleen A. Yonda*
National Sales Manager *Douglas J. DiNardo*
Marketing Manager *Amy Halloran*
Advertising Manager *Janelle Keeffer*
Production Editorial Manager *Renée Menne*
Publishing Services Manager *Karen J. Slaght*
Royalty/Permissions Manager *Connie Allendorf*

Wm. C. Brown Communications, Inc.

President and Chief Executive Officer *G. Franklin Lewis*
Senior Vice President, Operations *James H. Higby*
Corporate Senior Vice President, President of WCB Manufacturing *Roger Meyer*
Corporate Senior Vice President and Chief Financial Officer *Robert Chesterman*

Cover design by National Graphics, Incorporated
Cover image © Tony Stone Images

Copyright © 1992, 1995 by Wm. C. Brown Communications, Inc.
All rights reserved

A Times Mirror Company

ISBN 0-697-15834-9

No part of this publication may be reproduced, stored in a retrieval
system, or transmitted, in any form or by any means, electronic,
mechanical, photocopying, recording, or otherwise, without the
prior written permission of the publisher.

Some of the laboratory experiments included in this text may be
hazardous if materials are handled improperly or if procedures are
conducted incorrectly. Safety precautions are necessary when you are
working with chemicals, glass test tubes, hot water baths, sharp
instruments, and the like, or for any procedures that generally require
caution. Your school may have set regulations regarding safety
procedures that your instructor will explain to you. Should you have any
problems with materials or procedures, please ask your instructor for
help.

Printed in the United States of America by Wm. C. Brown Communications, Inc.,
2460 Kerper Boulevard, Dubuque, IA 52001

Contents

Introduction vii
Acknowledgments viii
Instructor's Note: Computer Data Disk ix
Materials Required for Each Experiment x

Experiment . . . . **Page**

1. Graphing 1
2. Ratios 15
3. Motion 27
4. Force Table 43
5. Free Fall 55
6. The Pendulum 65
7. Projectile Motion 77
8. Newton's Second Law 87
9. Conservation of Momentum 99
10. Rotational Equilibrium 107
11. Centripetal Force 117
12. Archimedes' Principle 123
13. Hooke's Law 129
14. Young's Modulus 137
15. Friction 147
16. Work and Power 157
17. Thermometer 161
18. Absolute Zero 169
19. Coeffient of Linear Expansion 179

20.	Specific Heat	183
21.	Static Electricity	191
22.	Electric Circuits	197
23.	Ohm's Law	203
24.	Series and Parallel Circuits	213
25.	Magnetic Fields	221
26.	Electromagnets	225
27.	Speed of Sound	229
28.	Standing Waves	237
29.	Reflection and Refraction	245
30.	Lenses	251
31.	Water Vapor	259
32.	Boyle's Law	265
33.	Nuclear Radiation	273

Appendix . . . . Page

I.	The Simple Line Graph	285
II.	The Slope of a Straight Line	287
III.	Experimental Error	289
IV.	Significant Figures	293
V.	Conversion of Units	297
VI.	Scientific Notation	299

Introduction

● This is a laboratory guide for a conceptual physics course. The purpose of the laboratory is to provide a hands-on introduction to experimental methods of scientific investigation, which will require the use of measurement and some mathematics skills such as graphing. An extensive mathematical background is not required in this laboratory guide, which will still provide a practical opportunity for you to gain a well-rounded understanding of physics. You will learn the theory and scientific laws pertaining to physical phenomena in the lecture portion of the course. Conducting experiments and collecting data to test the validity of theories and laws requires a different set of skills than those required for success in the lecture part of the course. Success in the laboratory involves skills in making accurate measurements of physical quantities in the real world, then formulating valid generalizations and principles based on the data. Every experiment in the laboratory will provide lessons and opportunities to learn these skills.

The instructions for each experiment in this guide include some basic theory and relationships about the physical phenomena to be investigated. These should be understood *before coming to the laboratory*. Do not waste valuable laboratory time trying to learn the theory and trying to figure out what to do.

● All measurements and data should be recorded in the data tables provided in this guide. Measurements should be made to provide full information, reading to the smallest scale division on the instrument, then estimating the last figure of the reading as closely as possible. See appendix IV, "Significant Figures," on page 293 for further information on making and recording measurement data.

Always bring your calculator and this laboratory guide to each session, completing required calculations and your laboratory report as you work. Be as accurate and neat as possible but do not waste time on the report. Sloppy work is to be avoided, of course, but *concentrate on knowing what you are doing, gathering accurate data, solving problems, and working out conclusions* while in the laboratory. These are the criteria for evaluating laboratory reports, not how much time you spend detailing a handsome report. Focus on these criteria during each laboratory session.

Most laboratory instructors expect a short, concise laboratory report submitted soon after each laboratory session. Your laboratory instructor will specify the exact format of your laboratory report. Two of the most important parts of a report are (1) the data and observations and (2) the conclusion. Note this laboratory manual is designed so that the data, the manipulations of data, and the conclusion can be recorded directly in the manual. The pages are perforated so they can be easily removed and submitted as individual laboratory reports.

● Data and observations include original data and observations that were gathered (recorded in tabular form when possible) as well as any manipulations of that data. Most of the manipulations will involve making some kind of graph to show a relationship between two measurements. In general, experiments require a slope of the graph that is compared to some value derived from theory. It is very important to remember when finding a slope of a graph to *avoid using data points* in your calculations. Two points whose coordinates are easy to find should be used instead of data points. One of the main reasons for plotting a graph and drawing a "best-fit" straight line is to smooth out any measurement errors made. Using data points directly in calculations defeats this purpose. To show what is happening as clearly as possible, all graphs should be a full page. There are helpful guides in the appendices on procedures and rules if you need help graphing or finding a slope.

The conclusion should state if you accomplished the purpose of the lab. This means accomplishing the purpose within a reasonable error, reasoning what constitutes a reasonable error, and stating what the origin of this error might be. In most of the labs, the purpose is to verify a law or theory that you have covered in class. Therefore, the conclusion may be as simple as "Newton's third law of motion was verified within 10% of the expected value." This should be followed by a statement as to why 10% is an acceptable error for this particular experiment. The conclusions should be reasonable and make sense, not necessarily agreeing with "expected" findings. Thoughtful analysis and careful, thorough thinking are much more important (and reasonable) than 0% error.

You are encouraged to work together in study groups, but your work should be your own. Note all the appendices at the end of this guide. They cover most of the problems that past students have had with laboratory skills and procedures.

Acknowledgments

The second edition of *Laboratory Manual in Conceptual Physics* reflects improvements resulting from requests for additional laboratory investigations and from suggested changes by users of the first edition. We were able to include most of the suggested new laboratory experiments that were requested, and the comments were very helpful in making improvements. Two short motion labs were combined into one single lab in this edition, and four new investigations were added.

For their special contributions to the second edition, I would like to acknowledge Howard Brown, Richland Community College; John Harper, Angelina College; Robert Hudson, Roanoke College. In addition, Tim Cook, Senior Laboratory Coordinator of the Department of Physics and Astronomy at Arizona State University deserves special mention for his many helpful suggestions and developmental testing of some of the procedures used in this manual.

Instructor's Note: Computer Data Disk

There is an available data disk for IBM PC and compatible or Macintosh computers, which will allow your students to experience some of the number crunching and graphing capabilities of a spreadsheet environment. The spreadsheets for our laboratory experiments were suggested, developed, and classroom tested by instructors like you. The disk is still under construction and is being provided free by Wm. C. Brown Publishers, along with an invitation for you to test the spreadsheets, suggest improvements, and submit new spreadsheet ideas.

The terminology of a spreadsheet is business oriented, and may at first seem unusual to those of us in other fields of study. Nonetheless, students can soon learn to use a spreadsheet as a "power calculator" to find answers to physics problems, and to display a graph of the results. Our use of the spreadsheet is more as a teaching tool rather than a calculator. We have structured the spreadsheet to give students an opportunity to extend a laboratory experiment, become more involved in the thought processes and the meaning of results, and to explore ideas. One of the early lab activities in this manual is concerned with teaching the skill of constructing and interpreting a graph. The related spreadsheet allows the student to explore variations of data on their own, seeing virtually instantaneous changes in the slope as they vary the data. In this way they might make some connections mentally about the meaning of a slope much faster than they would otherwise. The spreadsheet will also allow "creative explorations" that would otherwise not be available. For example, one of the lab activities in this manual is concerned with measuring the acceleration of gravity and free fall. The related spreadsheet allows the student to change the value of g, exploring how the object might fall on the moon, for example.

Spreadsheet applications are currently available for the following experiments:
- 1. Graphing (page 1)
- 2. Ratios (page 15)
- 3. Motion (page 27)
- 5. Free Fall (page 55)
- 7. Projectile Motion (page 77)
- 8. Newton's Second Law (page 87)

All of these spreadsheets are concerned with teaching some of the introductory skills needed in the early laboratory experiments, such as graphing, calculating a slope, and calculating an experimental error. Each spreadsheet is intended to be used *after the laboratory work for each experiment is completed* but before a lab report is submitted. Many of these spreadsheets provide a hypothetical data table, with instructions for the student to manipulate and graph the data (real data can be used later). The student is involved in a "real time" process as a plastic strip is placed on the screen (held in place by static electricity), so the student can easily move it about to see different results. The spreadsheets are protected, saved in a way that allows students to place data only in certain data cells, then enter the data to see the results in a graph. Some have the origin protected with 0,0 already entered, designed to cause many students to ask you questions. The typical nonscience student is not threatened by this "friendly" computer application, and can quickly learn to use the spreadsheet as a learning tool.

Using the spreadsheet as an exploratory teaching tool does not require a room full of computers. One or two can be set up in a laboratory so students can use them when they have time available. We hope you will be able to use our teaching tool spreadsheets and involve students in laboratory work from a different perspective. We would like very much to hear about your experiences, your ideas, and perhaps see a contribution with your name added to our Data Disk—under construction.

Materials Required for Each Experiment
(Quantities given are for an individual or teams of students.)

1. **Graphing** (page 1): Meterstick, masking tape, several types of small balls that bounce (tennis, ping-pong, rubber hand ball, etc).

2. **Ratios** (page 15): Several sizes of cups or beakers with round bottoms for tracing circles, metric rulers, string, wood or plastic cubes (about 30), three sizes of rectangular containers to hold water, balance, graduated cylinder.

3. **Motion** (page 27): Battery-operated toy bulldozer (or other toy car), long sheets of computer paper, butcher paper, or adding machine paper; masking tape, meterstick, stopwatch, inclined ramp (1 m or longer), 1 to 2 m rolling ball track, steel ball or marble (alternate setup: lab cart, photogates, computer with timer software).

4. **Force Table** (page 43): Force table with pulleys, string, and mass hangers; protractor and ruler.

5. **Free Fall** (page 55): Free-fall apparatus with spark timer, mass, meterstick, metric ruler (alternate setup: lab cart on track, photogate and pulley with spokes, computer with timer software).

6. **The Pendulum** (page 65): Different masses for bobs, support for pendulum, light cord or nylon, meterstick, stopwatch.

7. **Projectile Motion** (page 77): Ramp for table top, ring stand and ring, small rubber ball, meterstick, stopwatch, Styrofoam cup.

8. **Newton's Second Law** (page 87): Air track with cart, spark timer (alternate setup: lab cart on track, photogate and pulley with spokes, computer with timer software), pulley, nylon string, mass hanger, five 50 g masses, metric ruler or meterstick.

9. **Conservation of Momentum** (page 99): Air track with two carts, spark timer or stopwatch (alternate setup: lab cart on track, photogate and pulley with spokes, computer with timer software), mass set, metric ruler or meterstick.

10. **Rotational Equilibrium** (page 107): Meterstick with knife-edge clamp and support stand, three movable mass hangers and meterstick clamps, mass set.

11. **Centripetal Force** (page 117): Mass hanger and small masses, nylon string, rubber stopper, holder (wood or plastic rod with small hole for string; see figure 11.1).

12. **Archimedes' Principle** (page 123): Laboratory balance, object such as large metal mass, overflow can and catch bucket, spring scale (or laboratory balance attached to ring stand), block of wood, ball of clay.

13. **Hooke's Law** (page 129): Laboratory spring, support for spring, weight hanger, masses for weight hanger, meterstick.

14. **Young's Modulus** (page 137): Young's modulus apparatus, optical lever, low-power laser, steel wire, 10 1-kg masses, meterstick, micrometer caliper, metric ruler, strip of white paper (such as computer paper), vernier caliper.

15. **Friction** (page 147): Laboratory balance, wood block, clean and dry board, pulley and support, light cord, weight hanger and masses, set of laboratory masses, including a 0.5 kg mass.

16. **Work and Power** (page 157): Stairwell, meterstick, stopwatch, and a scale for weighing people.

17. **Thermometer Fixed Points** (page 161): Laboratory thermometer, beakers, crushed ice, steam generator, laboratory mercury barometer.

18. **Absolute Zero** (page 169): Apparatus (see fig. 18.1 on page 170) of large glass tube fitted with 2-hole rubber stoppers, small glass tube longer than large tube, two short glass tubes, rubber tubing and clamp, meterstick, mercury, thermometer, large beaker, funnel, crushed ice, hot water, steam generator.

19. **Coefficient of Linear Expansion** (page 179): Linear expansion laboratory apparatus, copper, aluminum, and iron rods or tubes to work with apparatus, steam boiler, heat source, thermometer, meterstick.

20. **Specific Heat** (page 183): Two Styrofoam cups to serve as a calorimeter, balance, three samples of shot made of different metals (e.g., aluminum, copper, lead), heating source for boiling water (hot plate and beaker or burner and ring stand setup), thermometer.

21. **Static Electricity** (page 191): Two glass rods, two hard rubber rods, nylon or silk cloth, wool cloth or fur, thread, electroscope, two rubber balloons.

22. **Electric Circuits** (page 197): Two flashlight batteries, tape or two flashlight battery holders, two flashlight bulbs, two flashlight bulb holders, hookup wire.

23. **Ohm's Law** (page 203): Adjustable dc power supply, dc voltmeter, dc ammeter, resistors, hookup wire or patch cords.

24. **Series and Parallel Circuits** (page 213): Three #41 bulbs (0.5 A), bulb sockets, two dry cells (or other source of 3 V dc), a dc voltmeter (0 to 5 volt range), and a dc ammeter (0 to 2 amp range), switch, hookup wire or patch cords.

25. **Magnetic Fields** (page 221): Large sheet of unlined white paper, small magnetic compass, bar

magnet, sharp pencil, large plastic sheet or glass plate, iron filings.

26. **Electromagnets** (page 225): Galvanoscope, small magnetic compass, dry cell or laboratory source of 1.5 V dc, about 3 m of No. 18 copper wire, 1/2 cm diameter soft iron spike about 12 cm long, box of regular-size paper clips.

27. **Speed of Sound** (page 229): Resonance tube apparatus, meterstick, two tuning forks of different pitches, rubber hammer, thermometer.

28. **Standing Waves** (page 237): Nylon string of known density, vibrator, pulley, mass hanger and small masses, meterstick, adjustable stroboscope (optional).

29. **Reflection and Refraction** (page 245): Ruler, cardboard (from box), small flat mirror, small wood block, rubber bands, straight pins, unlined white paper, protractor, 5 cm square glass plate.

30. **Lenses** (page 251): Convex lenses (long and short focal length), lens holders, meterstick, meterstick supports, tagboard screen, holder for tagboard screen, luminous object (clear glass light bulb or candle).

31. **Water Vapor** (page 259): Sling psychrometer (or two thermometers with a 3 cm length of cotton shoelace on the bulb end of one), meterstick.

32. **Boyle's Law** (page 265): Boyle's law apparatus, barometer.

33. **Nuclear Radiation** (page 273): Geiger counter, radioactive source (gamma emitting), meterstick, sheets of lead foil, two ring stands and ring stand clamps, alpha and beta emitting radioactive sources (optional), materials with various attenuation properties (optional).

Name_____Section_____Date_____

Experiment 1: Graphing

Introduction

The purpose of this introductory laboratory exercise is to gain experience in gathering and displaying data from a simple experiment. Refer to figure 1.1 for terminology used when discussing a graph and see appendix I for a detailed discussion about the terms.

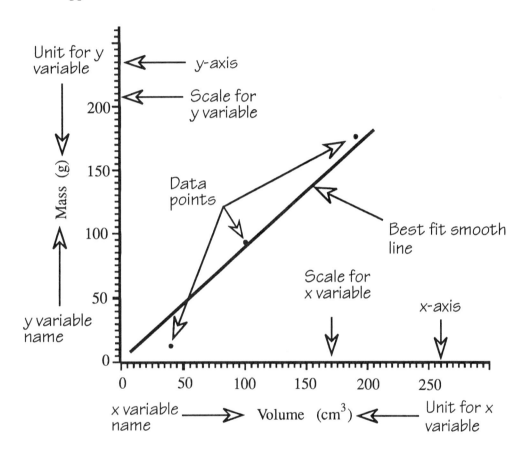

Figure 1.1

Procedure

1. Position a meterstick vertically on a flat surface, such as a wall or the side of a lab bench. Be sure the metric scale of the meterstick is on the outside and secure the meterstick to the wall or lab bench with two strips of masking tape.

2. Drop a ball as close as possible to the meterstick and measure (a) the height dropped, and (b) the resulting height bounced. Repeat this for three different heights dropped and record all data in Data Table 1.1 on page 5. In the data table, identify the *independent* (manipulated) variable and the *dependent* (responding) variable.

1

3. Use the graph paper on page 9 to make a graph of the data in Data Table 1.1, being sure to follow all the rules of graphing (see appendix I on page 285 for help). Title the graph, "Single Measurement Bounce Height as a Function of Height Dropped."

4. After constructing the graph, but before continuing with this laboratory investigation, answer the following questions:

   (a) What decisions did you have to make about how you conducted the ball-dropping investigation?

   (b) Would you obtain the exact same result if you dropped the ball from the same height several times? Explain.

   (c) Did you make a dot–to–dot line connecting the data points on your graph? Why or why not?

   (d) Could you use your graph to obtain a predictable result for dropping the ball from different heights? Explain why you could or could not.

(e) What is the significance of the origin on the graph of this data? Did you use the origin as a data point? Why or why not?

5. Make at least three more measurements for each of the previous three *height-dropped* levels. Find the average *height bounced* for each level and record the data and the average values in Data Table 1.2 on page 5.

6. Make a new graph of the *average height bounced* for each level that the ball was dropped. Draw a *straight best fit line* that *includes the origin* by considering the general trend of the data points. Draw the straight line as close as possible to as many data points as you can. Try to have about the same number of data points on both sides of the straight line. Title this graph, "Averaged Bounce Height."

7. Compare how well both graphs, "Single Measurement Bounce Height" and "Averaged Bounce Height," predict the heights that the ball will bounce for *heights dropped* that were not tried previously. Locate an untried height-dropped distance on the straight line, then use the corresponding value on the scale for height bounced as a prediction. Test predictions by noting several different heights, then measuring the actual heights bounced. Record your predictions and the actual experimental results in Data Table 1.3 on page 6.

8. Use a new, *different kind of ball* and investigate the bounce of this different ball. Record your single-measurement data for this different ball in Data Table 1.4. Record the averaged data for the height of the bounce for the three levels of dropping in Data Table 1.5. Repeat procedure step 7 for the different kind of ball. Record your predictions and the actual results in Data Table 1.6 on page 7.

9. Graph the results of the different kind of ball investigations onto the two previous graphs. Be sure to distinguish between sets of data points and lines by using different kind of marks. Explain the meaning of the different marks in a *key* on the graph.

10. What does the steepness (slope) of the lines tell you about the bounce of the different balls?

Results

1. Describe the possible sources of error in this experiment.

2. Describe at least one way that data concerning two variables is modified to reduce errors in order to show general trends or patterns.

3. How is a graph modified to show the best approximation of theoretical, error-free relationships between two variables?

4. Compare the usefulness of a graph showing (a) exact, precise data points connected dot–to–dot and (b) an approximated straight line that has about the same number of data points on both sides of the line.

5. Was the purpose of this lab accomplished? Why or why not? (Your answer to this question should be reasonable and make sense, showing thoughtful analysis and careful, thorough thinking.)

CUSTOMIZED LAB MANUAL ORDER FORM

TILLERY: *LABORATORY MANUAL IN CONCEPTUAL PHYSICS*

Professor _____

School address _____

School _____

City _____

State _____

Fall Order _____ Spring Order _____ Desk Copies _____ Total Order _____

Course Start Date _____

Please complete the top part of this form. Then number the exercises in the order that you would like them to appear in your customized manual. Preface and Appendix materials are included at no extra charge.

_____ Graphing (Ex. 1) _____ Absolute Zero (Ex. 17)

_____ Ratios (Ex. 2) _____ Coefficient of Linear Expansion (Ex. 18)

_____ Motion (Ex. 3) _____ Specific Heat (Ex. 19)

_____ Force Table (Ex. 4) _____ Static Electricity (Ex. 20)

_____ Free Fall (Ex. 5) _____ Electric Circuits (Ex. 21)

_____ Pendulum (Ex. 6) _____ Ohm's Law (Ex. 22)

_____ Projectile Motion (Ex. 7) _____ Series and Parallel Circuits (Ex. 23)

_____ Newton's Second Law (Ex. 8) _____ Magnetic Fields (Ex. 24)

_____ Conservation of Momentum (Ex. 9) _____ Electromagnetic (Ex. 25)

_____ Rotational Equilibrium (Ex. 10) _____ Speed of Sound (Ex. 26)

_____ Centripetal Force (Ex. 11) _____ Standing Waves (Ex. 27)

_____ Archimedes' Principle (Ex. 12) _____ Reflection and Refraction (Ex. 28)

_____ Hooke's Law (Ex. 13) _____ Lenses (Ex. 29)

_____ Young's Modules Friction (Ex. 14) _____ Water Vapor (Ex. 30)

_____ Work and Power (Ex. 15) _____ Boyle's Law (Ex. 31)

_____ Thermometer Fixed Points (Ex. 16) _____ Nuclear Radiation (Ex. 32)

Appendix:
    The Simple Line Graph
    The Slope of a Straight Line
    Experimental Error
    Significant Figures
    Conversion of Units
    Scientific Notation

Total number of exercises ordered _____

Extra material ☐ yes ☐ no

Special instructions _____

_____

_____

_____

_____

_____

_____

_____

_____

Please note: It takes 3-4 weeks to manufacture your customized text. To guarantee prompt delivery to your bookstore, we ask that Fall orders be placed no later than June 1; Spring orders by November 1; and Summer orders by April 15.

For total cost, ask your sales representative or call the CourseWorks Department at 1-800-228-0634.

Please return the completed order form to:

    Project Coordinator at C200
    Wm. C. Brown Publishers
    Customized Services
    2460 Kerper Blvd.
    Dubuque, IA 52001

spreadsheet software

This software is designed to teach and reinforce important concepts and skills needed for physics laboratory work. Each of the six spreadsheet activities is designed for a specific laboratory activity but builds on the skills learned in the previous activities. Each spreadsheet uses an activity to build critical skills that will be important in understanding most of the upcoming laboratory activities. Note the spreadsheets are used after the measurement part of a laboratory session is done, so one or two computers in the lab room should be sufficient.

instructor

To receive a complimentary copy of software supplementing labs 1, 2, 3, 5, 7, and 8, call **1-800-228-0459** and request:

       ISBN 0-697-26355-X   Windows
       ISBN 0-697-26356-8   Macintosh

student

If you are interested in using this software, request information from your professor or the bookstore manager. They may have ordered the spreadsheets for you already!

Data Table 1.1   Single Measurement Data: 1st Ball

Trial	Height Dropped _____variable	Height Bounced _____variable
1	_____	_____
2	_____	_____
3	_____	_____

Data Table 1.2   Averaged Bounce Data: 1st Ball

| Dropped Height | Bounce Height | | | Average |
	Trial 1	Trial 2	Trial 3	
_____	_____	_____	_____	_____
_____	_____	_____	_____	_____
_____	_____	_____	_____	_____

Data Table 1.3	Predictions and Results: 1st Ball					
Trial	Single Measurement Data			Averaged Data		
	Dropped Height	Predicted Height	Measured Height	Dropped Height	Predicted Height	Measured Height
1	_____	_____	_____	_____	_____	_____
2	_____	_____	_____	_____	_____	_____
3	_____	_____	_____	_____	_____	_____

Data Table 1.4	Single Measurement Data: 2nd Ball	
Trial	Height Dropped	Height Bounced
	_____variable	_____variable
1	_____	_____
2	_____	_____
3	_____	_____

Data Table 1.5   Averaged Bounce Data: 2nd Ball

Dropped Height	Bounce Height			Average
	Trial 1	Trial 2	Trial 3	
_____	_____	_____	_____	_____
_____	_____	_____	_____	_____
_____	_____	_____	_____	_____

Data Table 1.6   Predictions and Results: 2nd Ball

Trial	Single Measurement Data			Averaged Data		
	Dropped Height	Predicted Height	Measured Height	Dropped Height	Predicted Height	Measured Height
1	_____	_____	_____	_____	_____	_____
2	_____	_____	_____	_____	_____	_____
3	_____	_____	_____	_____	_____	_____

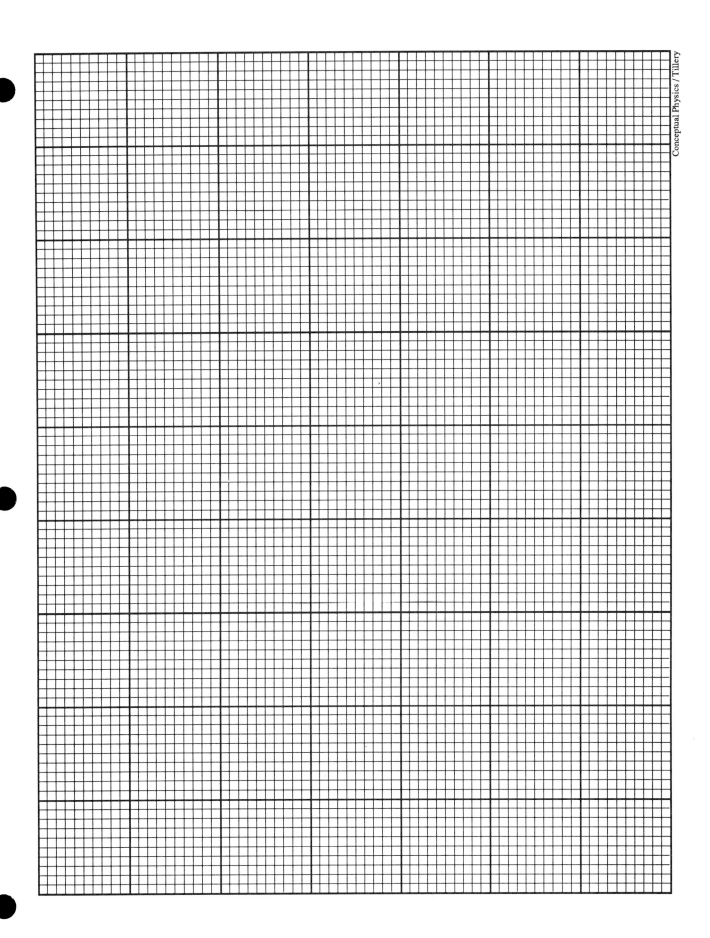

Name_____Section_____Date_____

Experiment 2: Ratios

Introduction

The purpose of this introductory laboratory exercise is to investigate how measurement data are simplified in order to generalize and identify trends in the data. Data concerning two quantities will be compared as a **ratio**, which is generally defined as a relationship between numbers or quantities. A ratio is usually simplified by dividing one number by another.

Procedure

Part A: Circles and Proportionality Constants

1. Obtain three different sizes of cups, containers, or beakers with round bottoms. Trace around the bottoms to make three large but different-sized circles on a blank sheet of paper (fig. 2.1).

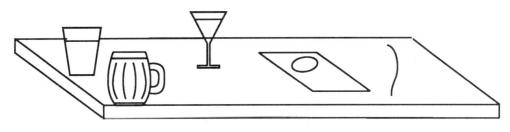

Figure 2.1

2. Mark the diameter on each circle by drawing a straight line across the center at the widest part. Measure each diameter in mm and record the measurements in Data Table 2.1. Repeat this procedure for each circle for a total of three trials.

3. Measure the circumference of each object by carefully positioning a length of string around the object, then grasping the place where the string ends meet. Measure the length in mm and record the measurements for each circle in Data Table 2.1. Repeat the procedure for each circle for a total of three trials. Find the ratio of the circumference of each circle to its diameter. Record the ratio for each trial in Data Table 2.1 on page 22.

4. The ratio of the circumference of a circle to its diameter is known as **pi** (symbol π), which has a value of 3.14 ... (the ellipses mean many decimal places). Average all the values of π in Data Table 2.1 and calculate the experimental error.

Part B: Area and Volume Ratios

1. Obtain one cube from the supply of same-sized cubes in the laboratory. Note that a cube has six sides, or six units of surface area. The side of a cube is also called a *face*, so each cube has six identical faces with the same area. The overall surface area of a cube can be found by measuring the length and width of one face (which should have the same value) and then multiplying (length)(width)(number of faces). Use a metric ruler to measure the cube, then calculate the overall surface area and record your finding for this small cube in Data Table 2.2 on page 22.

2. The volume of a cube can be found by multiplying the (length)(width)(height). Measure and calculate the volume of the cube and record your finding for this small cube in Data Table 2.2.

3. Calculate the ratio of surface area to volume and record it in Data Table 2.2.

4. Build a medium-sized cube from eight of the small cubes stacked into one solid cube. Find and record (a) the overall surface area, (b) the volume, and (c) the overall surface area to volume ratio, and record them in Data Table 2.2.

5. Build a large cube from 27 of the small cubes stacked into one solid cube. Again, find and record the overall surface area, volume, and overall surface area to volume ratio and record your findings in Data Table 2.2.

6. Describe a pattern, or generalization, concerning the volume of a cube and its surface area to volume ratio. For example, as the volume of a cube increases, what happens to the surface area to volume ratio? How do these two quantities change together for larger and larger cubes?

Part C: Mass and Volume

1. Obtain at least three straight-sided, rectangular containers. Measure the length, width, and height *inside* the container (you do not want the container material included in the volume) as shown in figure 2.2. Record these measurements in Data Table 2.3 (page 23) in rows 1, 2, and 3. Calculate and record the volume of each container in row 4 of the data table.

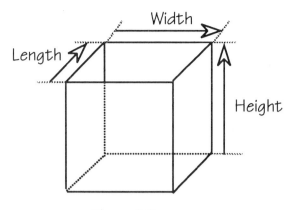

Figure 2.2

2. Measure and record the mass of each container in row 5 of the data table. Measure and record the mass of each container when "level full" of tap water. Record each mass in row 6 of the data table. Calculate and record the mass of the water in each container (mass of container plus water minus mass of empty container, or row 6 minus row 5 for each container). Record the mass of the water in row 7 of the data table.

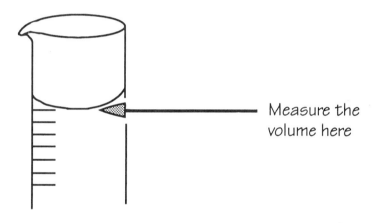

Figure 2.3

3. Use a graduated cylinder to measure the volume of water in each of the three containers (see fig. 2.3). Be sure to get *all* the water into the graduated cylinder. Record the water volume of each container in milliliters (mL) in row 8 of the data table.

4. Calculate the ratio of cubic centimeters (cm^3) to mL for each container by dividing the volume in cubic centimeters (row 4 data) by the volume in milliliters (row 8 data). Record your findings in the data table.

5. Calculate the ratio of mass per unit volume for each container by dividing the mass in grams (row 7 data) by the volume in milliliters (row 8 data). Record your results in the data table.

6. Make a graph of the mass in grams (row 7 data) and the volume in milliliters (row 8 data) to picture the mass per unit volume ratio found in step 5. Put the volume on the *x*-axis (horizontal axis) and the mass on the *y*-axis (the vertical axis). The mass and volume data from each container will be a data point, so there will be a total of three data points.

7. Draw a straight line on your graph that is as close as possible to the three data points and the origin (0, 0) as a fourth point. If you wonder why (0, 0) is also a data point, ask yourself about the mass of a zero volume of water!

8. Calculate the slope of your graph. (See appendix II on page 287 for information on calculating a slope.)

9. Calculate your experimental error. Use 1.0 g/mL (grams per milliliter) as the accepted value.

10. Density is defined as mass per unit volume, or mass/volume. The slope of a straight line is also a ratio, defined as the ratio of the change in the *y*-value per the change in the *x*-value. Discuss why the volume data was placed on the *x*-axis and mass on the *y*-axis and not vice versa.

11. Was the purpose of this lab accomplished? Why or why not? (Your answer to this question should show thoughtful analysis and careful, thorough thinking.)

Results

1. What is a ratio? Give several examples of ratios in everyday use.

2. How is the value of π obtained? Why does π not have units?

3. Describe what happens to the surface area to volume ratio for larger and larger cubes. Predict if this pattern would also be observed for other geometric shapes such as a sphere. Explain the reasoning behind your prediction.

4. Why does crushed ice melt faster than the same amount of ice in a single block?

5. Which contains more potato skins: 10 pounds of small potatoes or 10 pounds of large potatoes? Explain the reasoning behind your answer in terms of this laboratory investigation.

6. Using your own words, explain the meaning of the slope of a straight-line graph. What does it tell you about the two graphed quantities?

7. Explain why a slope of mass/volume of a particular substance also identifies the density of that substance.

Problems

An aluminum block that is 1 m × 2 m × 3 m has a mass of 1.62×10^4 kilograms (kg). The following problems concern this aluminum block:

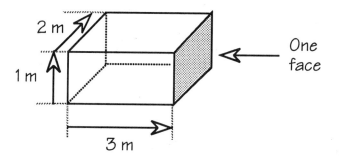

Figure 2.4

1. What is the volume of the block in cubic meters (m³)?

2. What are the dimensions of the block in centimeters (cm)?

3. Make a sketch of the aluminum block and show the area of each face in square centimeters (cm²).

4. What is the volume of the block expressed in cubic centimeters (cm³)?

5. What is the mass of the block expressed in grams (g)?

6. What is the ratio of mass (g) to volume (cm³) for aluminum?

7. Under what topic would you look in the index of a reference book to check your answer to question 6? Explain.

Data Table 2.1 Circles and Ratios

	Small Circle			Medium Circle			Large Circle		
Trial	1	2	3	1	2	3	1	2	3
Diameter (D)	___	___	___	___	___	___	___	___	___
Circumference (C)	___	___	___	___	___	___	___	___	___
Ratio of C/D	___	___	___	___	___	___	___	___	___

Average $\frac{C}{D}$ = _____   Experimental error: _____

Data Table 2.2 Area and Volume Ratios

	Small Cube	Medium Cube	Large Cube
Surface Area	_____	_____	_____
Volume	_____	_____	_____
Ratio of Area/Volume	_____	_____	_____

Data Table 2.3	Mass and Volume Ratios		
Container Number	1	2	3
1. Length of container	_____cm	_____cm	_____cm
2. Width of container	_____cm	_____cm	_____cm
3. Height of container	_____cm	_____cm	_____cm
4. Calculated volume	_____cm^3	_____cm^3	_____cm^3
5. Mass of container	_____g	_____g	_____g
6. Mass of container and water	_____g	_____g	_____g
7. Mass of water	_____g	_____g	_____g
8. Measured volume of water	_____mL	_____mL	_____mL
9. Ratio of calculated volume to measured volume of water	_____cm^3/mL	_____cm^3/mL	_____cm^3/mL
10. Ratio of mass of water to measured volume of water	_____g/mL	_____g/mL	_____g/mL

Experiment 3: Motion

Introduction

In this investigation you will analyze and describe motion with a constant velocity and motion with a nonconstant velocity. First, motion with a constant velocity will be investigated by using a battery-operated toy bulldozer, or any toy car or truck that moves at a fairly constant speed. Data will be collected, analyzed, and a concept will be formalized to describe what is happening to the toy as it moves.

Figure 3.1 compares the distance versus time slopes for motion with a constant velocity, with a nonconstant velocity, and with no velocity at all. Note that the slope for some object not moving will be a straight horizontal line. If a vehicle is moving at a uniform (constant) velocity the line will have a positive slope. This slope will describe the magnitude of the velocity, sometimes referred to as the **speed**. The line for a vehicle moving at a nonconstant speed, on the other hand, will be nonconstant as shown in figure 3.1. A nonconstant speed is also known as accelerated motion, and the ratio of how fast the motion is changing per unit of time is called **acceleration**.

Taking measurable data from a multitude of sense impressions, finding order in the data, then inventing a concept to describe the order are the activities of science. This investigation applies this process to motion.

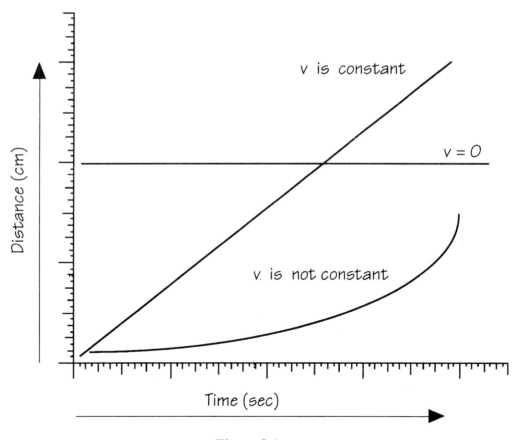

Figure 3.1

Procedure

Part A: Constant Velocity on the Level

1. Use masking tape to secure a length of paper such as long sheets of computer paper, rolled butcher paper, or adding machine tape across the floor. The paper should be long enough so the motorized toy vehicle used will not cross the entire length in less than 8 to 10 seconds. Thus, the exact length of paper selected will depend on the vehicle and battery conditions. (Note: Erratic increases or decreases of speed probably mean that a new battery is needed.) The paper will be used to record successive positions of the toy at specific time intervals as shown in figure 3.2.

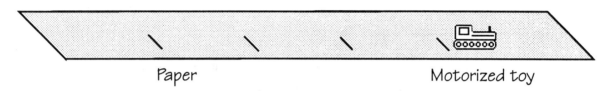

Figure 3.2

2. One person with a stopwatch will call out equal time intervals that are manageable but result in at least five or six data points for the total trip. Another person will mark the position of the toy vehicle on the paper when each time interval is called. To avoid interfering with the motion of the toy, mark the position from behind each time. This also means that the starting position should be marked from behind. Other means of measuring velocity that might be used in your laboratory, such as the use of photogates and computer software, will be explained by your instructor.

3. Measure the intervals between the time marks, recording your data in Data Table 3.1 on page 33. Make a graph that describes the motion of the toy vehicle by placing the distance (the dependent variable) on the vertical axis and time (the independent variable) on the horizontal axis. Draw the best straight line as close as possible to the data points. Calculate the slope and record it someplace on the graph.

Part B: Constant Velocity on an Incline

1. This investigation is similar to part A, but this time the toy tractor will move up an inclined ramp that is at least 1 m long.

2. Elevate the ramp with blocks or books so that 1 meter from the bottom of the ramp is 10 cm high. As in part A, one person with a stopwatch will call out equal time intervals that are manageable, but result in at least five or six data points for the total trip. Another person will mark the position of the toy vehicle on the paper when each time interval is called. To avoid interfering with the motion of the toy, mark the position from behind each time. Also mark the starting position from behind. Measure the intervals between the time marks, recording your data in Data Table 3.2.

3. Elevate the ramp to 20 cm high and repeat procedure step 2.

4. Elevate the ramp to 30 cm high and again repeat procedure step 2. Make a graph of all three sets of data in Data Table 3.2. Calculate the slope of each line and write each somewhere on the graph.

Part C: Motion with Nonconstant Velocity

1. You will now set up a track for collecting data about rolling balls. This track can be anything that serves as a smooth, straight guide for a rolling ball. It could be a board with a V-shaped groove, U-shaped aluminum shelf brackets, or two lengths of pipe taped together, for example.

2. The track should be between 1 and 2 m long and supported somewhere between 10 and 50 cm above the table at the elevated end (figure 3.3). In this investigation, a longer track will mean better results. You should consider 1 m as a *minimum* length. Your instructor will describe a different procedure if your lab has photogates, computer software, or different equipment.

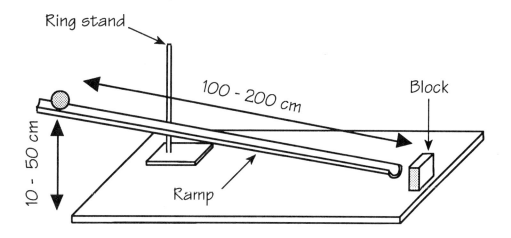

Figure 3.3

3. You will select a minimum of six positions on the ramp from which to release a steel ball or marble. One position should be the uppermost end and the others should be equally spaced. Hold a ruler across the track with the ball behind it, then release the ball by lifting the ruler straight up the same way each time. Start a stopwatch when the ball is released, then stop it when the ball reaches the bottom of the ramp. A block at the bottom of the ramp will stop the ball and the sound of the ball hitting the block will signal when to stop the stopwatch.

4. Measure the distance and time for three data runs, then average the data for each of the six positions. Record the data in Data Table 3.3 on page 35. Make a graph of the data with time on the *x*-axis.

Results

1. Explain for each part of this investigation how you know if there is or is not a relationship between the variables according to your graphs.

2. For motion with a constant velocity, how do the changes in distance compare for equal time intervals? Is this what you would expect? Explain.

3. What is the rate of travel of the toy over (a) a flat surface, (b) a surface elevated 10 cm high, (c) a surface elevated 20 cm high, and (d) a surface elevated 30 cm high?

4. For motion with a nonconstant velocity, how does the total distance change as the total time increases; that is, do they both increase at the same rate? Explain the meaning of this observation.

5. Considering nonconstant velocity, how do the changes in distance compare for equal time intervals?

6. Was the purpose of this lab accomplished? Why or why not? (Your answer to this question should show thoughtful analysis and careful, thorough thinking.)

Going Further

In part of this investigation, you learned that $\bar{v} = \frac{d}{t}$. Using this equation, explain how you can find

(a) the time for a trip when given the average speed and the total distance traveled;

(b) the total distance traveled when given the time for a trip and the average speed; and

(c) the average speed for a trip, no matter what units are used to describe the total distance and the total time of the trip.

Data Table 3.1   Distance and Time Data for Battery-Powered Toy over a Flat Surface

Total Time (s)	Total Distance (cm)
_____	_____
_____	_____
_____	_____
_____	_____
_____	_____
_____	_____

Data Table 3.2   Distance and Time Data for Battery-Powered Toy over Elevated Surfaces

Time (s)	Total Distance (cm)		
	10 cm Elevation	20 cm Elevation	30 cm Elevation
_____	_____	_____	_____
_____	_____	_____	_____
_____	_____	_____	_____
_____	_____	_____	_____
_____	_____	_____	_____

Data Table 3.3		Time and Distance Data for Rolling Ball on Ramp		
Distance from Bottom (cm)	Time Trial 1 (s)	Time Trial 2 (s)	Time Trial 3 (s)	Time Average (s)
_____	_____	_____	_____	_____
_____	_____	_____	_____	_____
_____	_____	_____	_____	_____
_____	_____	_____	_____	_____
_____	_____	_____	_____	_____
_____	_____	_____	_____	_____

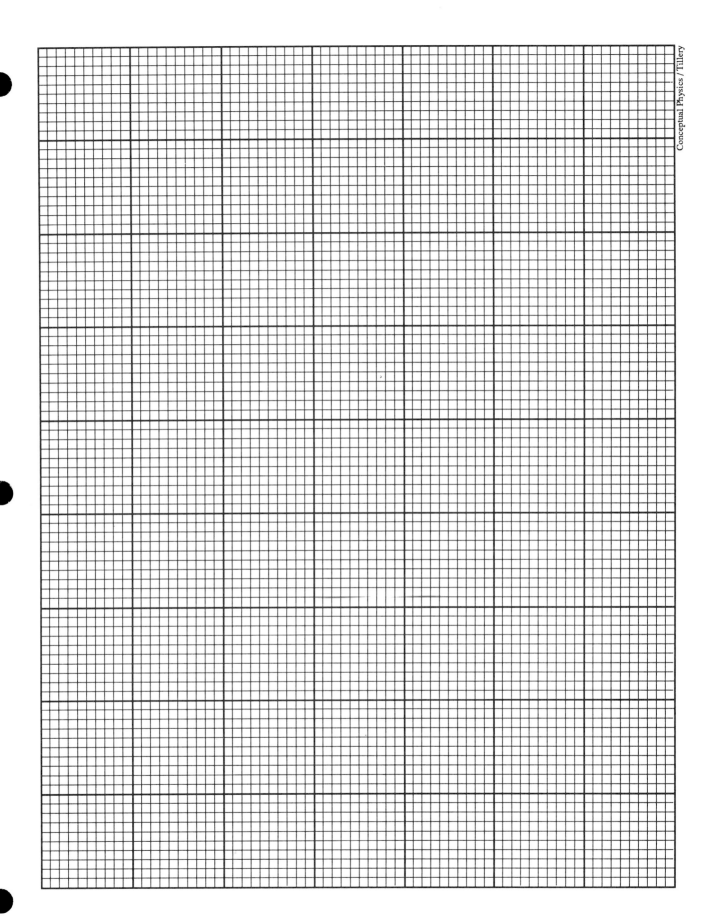

Name_____Section_____Date_____

Experiment 4: Force Table

Introduction

When you push on something, you know that you can vary the size, or *magnitude,* of the exerted force by varying your muscular exertion. You can also vary the *direction* that you push, exerting a force in practically any direction. For example, if you and another person push on a stalled car to start it moving, you know the direction you both push is as important as the magnitude of your push. If you are both side by side, pushing in the same direction, your forces add to some net force in the same direction. If you push in parallel but opposite directions, however, one force will cancel the other and the net force will be zero. In between the two extremes of parallel forces in the same direction and parallel forces in opposite directions, two people pushing at various directions will produce net forces somewhere between a net force of zero and a maximum net force from adding the two forces together. It should be clear that a force has both a magnitude, or size, and it has a direction. Quantities that have both magnitude and direction like this are called **vectors**. Velocity, acceleration, and a force all have magnitude and direction, so they are vector quantities. Quantities that have magnitude only are called **scalars**. Volume, temperature, and area all have magnitude only, so they are scalar quantities. Scalar quantities can be added by ordinary arithmetic, which is called scalar addition. Scalar addition will not work with vectors, however, since these quantities have a property of direction.

The vector sum of two or more vectors is called the **resultant**. There are two basic methods used to find a resultant: (1) the graphical method and (2) the analytical component method. In the graphical method a vector is illustrated by drawing a straight line with an arrowhead pointing in the direction of the vector. The length of the arrow is drawn to a scale to represent the magnitude of the vector quantity. For example, when two displacement vectors are drawn on the same diagram, the arrow representing the larger displacement must be longer than the arrow representing the shorter displacement. Figure 4.1 shows the resultant of 103 N when a force of 50 N is applied toward the North and a force of 90 N is applied to the East.

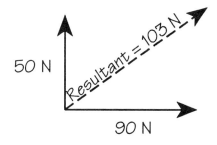

Figure 4.1

According to Newton's laws of motion, an object at rest with two or more forces acting on it must have a vector sum, or resultant, of zero. An object with two or more forces with a resultant of zero is said to be in **equilibrium**. In equilibrium there is a vector called the **equilibrant** that cancels the effects of the other vectors. The equilibrant always has the same magnitude but opposite direction as the resultant. In this experiment you will investigate resultants, equilibrants, and other force vectors in equilibrium, using the graphical method of vector addition.

Procedure

Examine the force table that you will use, which may be like the one illustrated in figure 4.2. Note how the pulleys can be moved around the table to various positions. A string is tied to a ring at the center, runs over a pulley, and down to a mass hanger. You can create forces of different magnitudes and directions on this ring by varying the mass on the hangers and by moving the pulleys.

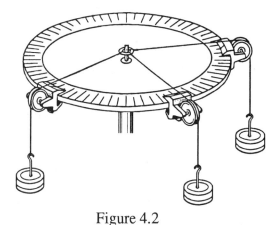

Figure 4.2

Part A: Adding Two Forces

1. Before starting make sure the force table is level, and adjust each string so it is level with the force table and aligned correctly on its pulley. Also check the string knot on the center ring to make sure the string is pointing to the center of the ring, not the edge or elsewhere. Check this from time to time and move the string knot as necessary to ensure alignment. Angles will be measured, then graphed, so orient the force table with the zero angle on the right side.

2. Place a total of 200 g (m_1) on one string (150 g mass plus the hanger, which is usually 50 g), which runs over a pulley attached to the force table at 0°. Place a total of 300 g on a second string (m_2), which runs over a pulley attached to the force table at 90°.

3. Record the masses of m_1 and m_2 in kilograms in Data Table 4.1 on page 49, including the mass of the hanger in each case. Calculate the force of gravity, and thus the force on each string: $\mathbf{F_1} = m_1 g$ and $\mathbf{F_2} = m_2 g$, where m is in kilograms and g is 9.80 m/s² acceleration of gravity. Record each force, in N, in Data Table 4.1.

4. Record in Data Table 4.1 the angle of each string from the positive x axis, measured from zero on the table scale on your right side.

5. On the top half of a sheet of graph paper, use a protractor and ruler to draw vector arrows for the force on each string on the force table, and label them F_1 and F_2. The protractor should be used to measure the angles from a positive x axis. Use a scale, such as 1 cm = 1 N, that will permit you to draw the longest arrows possible on the top half of the paper when the arrows are placed head to tail. Record the scale somewhere on the graph paper.

6. On the bottom half of the sheet of graph paper, use a ruler and protractor to carefully draw F_1 again, this time with its tail at the head of F_2. Draw the resultant, R, between the open ends of F_2 and F_1. Measure the length of R and determine the magnitude of the resultant in N by using the scale. Use the protractor to find the angle of R from the x axis. Record this graphical result of both the magnitude and angle of the resultant in Data Table 4.1.

7. Check your graphical result by attaching a third pulley to the force table. Attach the pulley at the angle that represents the equilibrant of R. Calculate the mass needed from m = F/g. Place this mass on the third string hanger, and remember to include the mass of the hanger itself. If the equilibrant balances the two forces, the ring will stay in the center without touching the post. You might want to jiggle the ring a little to make sure the system is not "stuck" by friction. Also make sure the length of each string is pointing to the center of the ring.

8. Check your graphical results again, this time by applying the Pythagorean theorem. As you can see in figure 4.3, in this situation the resultant R is equal to the square root of $(F_1)^2 + (F_2)^2$. Show your work below figure 4.3.

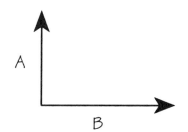

Two forces acting on a point.

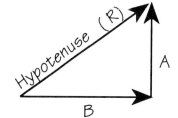

$R^2 = A^2 + B^2 \quad \therefore \quad R = \sqrt{A^2 + B^2}$

Figure 4.3

Calculations using the Pythagorean theorem:

Part B: Adding Three Forces

1. Arrange three pulleys at some angles, as shown by the vector arrows in figure 4.4. Do not record any angles at this time.

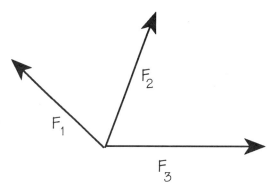

Figure 4.4

2. Add 300 to 600 g masses to each hanger, and identify the masses as m_1, m_2, and m_3 in Data Table 4.2 on page 50. Do not forget to include the mass of the hanger in the total for each string. Calculate the force of gravity, and thus the force on each string: $\mathbf{F_1} = m_1 g$, $\mathbf{F_2} = m_2 g$ and $\mathbf{F_3} = m_3 g$ where m is in kilograms and g is 9.80 m/s² acceleration of gravity. Record the magnitude of each force in Data Table 4.2.

3. Move one pulley, then another, in one degree intervals until the ring stays in the middle without touching the post. Jiggle the ring to make sure the system is not stuck because of friction. When the system is balanced, record in Data Table 4.2 the angles, with the zero of the table scale on your right side.

4. On the top half of a sheet of graph paper, use a protractor and ruler to carefully draw vector arrows for the force on each string on the force table, and label them $\mathbf{F_1}$, $\mathbf{F_2}$, and $\mathbf{F_3}$. A protractor should be used to measure the angles from the positive x axis. Use a scale, such as 1 cm = 1 N, that will permit you to draw the longest arrows possible on the top half of the paper when the arrows are placed head to tail. Record the scale somewhere on the graph paper.

5. On the bottom half of the sheet of graph paper, use a ruler and protractor to carefully draw arrows representing the vector forces. Draw $\mathbf{F_2}$ first, then draw $\mathbf{F_1}$ with its tail at the head of $\mathbf{F_2}$. Carefully draw $\mathbf{F_3}$ with its tail at the head of $\mathbf{F_2}$. Draw the resultant, $\mathbf{R}$, between the open ends of $\mathbf{F_1}$ and $\mathbf{F_3}$. Measure the length of $\mathbf{R}$ and determine the magnitude of the resultant in N by using the scale. Use the protractor to find the angle of $\mathbf{R}$ from the x axis. Record this graphical result of both the magnitude and angle of the resultant in Data Table 4.2.

6. Check your graphical result by attaching a fourth pulley to the force table. Attach the pulley at the angle that represents the equilibrant of $\mathbf{R}$. Calculate the mass needed from m = F/g. Place this mass on the third string hanger, and remember to include the mass of the hanger itself. If the equilibrant

balances the two forces, the ring will stay in the center without touching the post. You might want to jiggle the ring a little to make sure the system is not stuck by friction. Also make sure the length of each string is pointing toward the center of the ring.

Results

1. Compare the results of the graphical solution, the experimental solution, and the mathematical solution in both parts of this experiment.

2. What are the advantages and disadvantages of using the (a) graphical, (b) experimental and (c) mathematical methods for adding vectors?

3. Describe some possible sources of error in this experiment. How could you improve the experiment?

4. Was the purpose of this lab accomplished? Why or why not? (Your answer to this question should be reasonable and make sense, showing thoughtful analysis and careful, thorough thinking.)

Data Table 4.1	Adding Two Vector Forces		
	Mass (Kg)	Force (N)	Angle (degrees)
String 1	_____	_____	_____
String 2	_____	_____	_____
Graphical Results: Resultant of F_1 and F_2		_____	_____
Experimental Results: Balance of F_1 and F_2 with equilibrant	_____		
Mathematical Results: Pythagorean theorem answer	_____		

Data Table 4.2	Adding Three Vector Forces		
	Mass (Kg)	Force (N)	Angle (degrees)
String 1	_____	_____	_____
String 2	_____	_____	_____
String 3	_____	_____	_____
Graphical Results: Resultant of $\mathbf{F_1}$, $\mathbf{F_2}$, and $\mathbf{F_3}$		_____	_____

Experimental Results: Balance of $\mathbf{F_1}$, $\mathbf{F_2}$, and $\mathbf{F_3}$ with equilibrant _____

_____

_____

_____

Name_____ Section_____ Date_____

Experiment 5: Free Fall

Introduction

In this experiment you will calculate the acceleration of an object as it falls toward the earth's surface. An object in *free fall* moves toward the surface with a uniform accelerated motion due to gravity, *g*. The value of *g* varies with location on the surface of the earth, increasing with latitude to a maximum at the poles. The value of *g* also varies with elevation, decreasing with elevation at a certain latitude. The average, or standard, value of *g*, however, is usually accepted as 9.80 m/s² or 980 cm/s².

When you measure the total distance that an object moves during some period of time, you can calculate an average velocity. **Average velocity** is defined as

$$\bar{v} = \frac{\Delta d}{\Delta t}$$

where Δd is the total distance (final distance minus initial, or $d_f - d_i$) and Δt is the total time (final time minus initial, or $t_f - t_i$). In this experiment you will be measuring the velocity of an object that falls from an initial distance and time of zero, so $\Delta d = d_f - 0$ and $\Delta t = t_f - 0$. For the case of a falling object,

$$\bar{v} = \frac{\Delta d}{\Delta t} = \frac{d_f - d_i}{t_f - t_i} \quad \text{since } d_i = 0 \text{ and } t_i = 0 \quad \therefore \quad \bar{v} = \frac{d_f}{t_f}.$$

Thus you can calculate the average velocity of an object in free fall from the total distance traveled and the time of fall. When an object moves with a constant acceleration, you can also find the average velocity by adding the initial and final velocity and dividing by 2,

$$\bar{v} = \frac{v_f + v_i}{2}.$$

By substituting the other expression for average velocity, we have

$$\bar{v} = \frac{v_f + v_i}{2} \quad \text{and} \quad \bar{v} = \frac{d_f}{t_f} \quad \therefore \quad \frac{v_f + v_i}{2} = \frac{d_f}{t_f}.$$

Since the initial velocity of a dropped object is zero, then v_i is zero and we can solve for the final velocity of v_f, and

$$\frac{v_f}{2} = \frac{d_f}{t_f} \quad \therefore \quad v_f = \frac{2d_f}{t_f}.$$

In this experiment you will measure the distance a mass has fallen during recurring time intervals according to a timing device. This data will enable you to calculate the instantaneous velocity at known time intervals. Plotting the velocity versus the time, then finding the slope will provide an experimental value of *g*.

Procedure

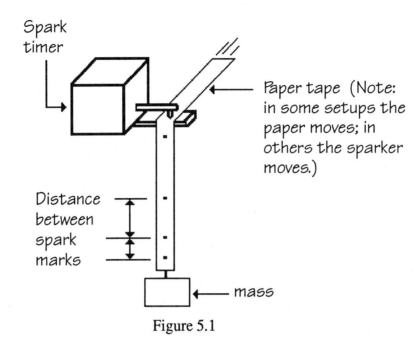

Figure 5.1

1. You will experimentally determine the acceleration due to gravity and compare it to the standard value of 980 cm/s^2. The procedure may vary with the apparatus used. For example, you might use an apparatus that consists of a device to measure the free fall of an object with a spark timer that will mark a paper tape at equal time intervals (fig. 5.1). As a mass accelerates downward it will leave a trail of spark marks at equal time intervals. You will draw a perpendicular line through each mark, then identify the first mark as your reference line. The first mark is identified as the place where $d_f = 0$. Other means of measuring velocity that might be used in your laboratory, such as the use of photogates and computer software, will be explained by your instructor.

2. For spark mark (or ink dot) trails measure the *total distance* (d_f) by using the beginning mark as a reference line. On page 59, record in Data Table 5.1 the distance in cm of each mark *from the reference line*.

3. For each spark, record the *time* (t) that elapsed between the marks as determined by the spark timer. Your instructor will provide exact information for your timer. Most timers are set to operate on 60 Hz, making a spark every 1/60 second. Thus the second spark would have occurred 1/60 second after the first, the third spark mark would have occurred 1/60 plus 1/60, or 2/60 (0.033 s) after the first mark. Fill in Data Table 5.1 with the total distance and time data for each mark, and calculate and record the velocity at each spark. Repeat the experiment two more times with two more paper tapes, completing Data Table 5.2 (page 60) and Data Table 5.3 (page 61).

Results

1. Look over the data in Data Tables 5.1, 5.2, and 5.3, think about what the date means, then select the Data Table that seems to have the "best run" data. State which table was chosen and explain the basis for your choice.

2. Using the data table from the best run, make a graph with *velocity* (v) on the y-axis and *elapsed time* (t) on the x-axis. (Note: Because the first spark was probably not made at the actual time of release, the line on your graph will probably not have a y-intercept of 0.) Find the slope and record it here, along with any notes you may wish to record.

3. Use the calculated slope and the accepted value of 980 cm/s^2 to calculate the experimental error.

4. Was the purpose of this lab accomplished? Why or why not? (Your answer to this question should show thoughtful analysis and careful, thorough thinking.)

Going Further

What is your reaction time? One way to measure your reaction time is to have another person hold a meterstick vertically from the top while you position your thumb and index finger at the 50 cm mark. The other person will drop the meterstick (unannounced) and you will catch it with your thumb and finger. Accelerated by gravity (g), the stick will fall a distance (d) during your reaction time (t). Knowing d and g, all you need is a relationship between g, d, and t to find the time.

You know a relationship between d, $\bar{v}$, and t from $\bar{v} = d/t$. Solving for d gives $d = \bar{v}t$.

Any object in free fall, including a meterstick, will have uniformly accelerated motion, so the average velocity is

$$\bar{v} = \frac{v_f + v_i}{2}.$$

Substituting for the average velocity in the previous equation gives

$$d = \left(\frac{v_f + v_i}{2}\right)(t).$$

The initial velocity of a falling object is always zero just as it is dropped, so the initial velocity can be eliminated, giving

$$d = \left(\frac{v_f}{2}\right)(t).$$

Now you want acceleration in place of velocity. From

$$\bar{a} = \frac{v_f - v_i}{t}$$

and solving for the final velocity gives

$$v_f = \bar{a}t.$$

The initial velocity is again dropped since it equals zero. Substituting the final velocity in the previous equation gives

$$d = \left(\frac{\bar{a}t}{2}\right)(t) \quad \text{or} \quad d = \frac{1}{2}\bar{a}t^2.$$

Finally, solving for t gives

$$t = \sqrt{\frac{2d}{g}}.$$

Measuring how far the meterstick falls (in m) can now be used as the distance (d) with g equaling 9.80 m/s^2 to calculate your reaction time (t).

Data Table 5.1   Free Fall Run Number One

Spark Number	Distance (cm)	Time of Fall (s)	Computed Instantaneous Velocity (cm/s)
1			
2			
3			
4			
5			
6			
7			
8			
9			
10			

Data Table 5.2	Free Fall Run Number Two		
Spark Number	Distance (cm)	Time of Fall (s)	Computed Instantaneous Velocity (cm/s)
1			
2			
3			
4			
5			
6			
7			
8			
9			
10			

Data Table 5.3	Free Fall Run Number Three		
Spark Number	Distance (cm)	Time of Fall (s)	Computed Instantaneous Velocity (cm/s)
1			
2			
3			
4			
5			
6			
7			
8			
9			
10			

Name_____Section_____Date_____

Experiment 6: The Pendulum

Introduction

A simple pendulum consists of a mass, called a **bob**, connected to the end of a suspended cord or string. When the bob is pulled to one side of its position of rest and then released, it begins to vibrate in a simple harmonic motion. In general, a **vibration** is any back-and-forth motion that repeats itself. Simple harmonic motion occurs when there is a restoring force that is equal and opposite to a displacement. The restoring force, of course, comes from gravity acting on the bob.

There are a number of factors that could affect the vibration of a pendulum and this investigation is concerned with these factors. A vibrating pendulum can be described by measuring several variables. The extent of displacement from the rest position is called the **amplitude** of the vibration (C to A – or B – in figure 6.1). A vibration that has the bob displaced a greater distance from the rest position thus has a greater amplitude than a vibration with less displacement.

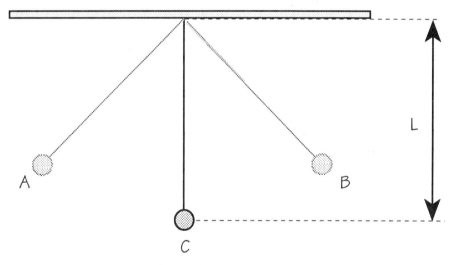

Figure 6.1

A complete vibration is called a cycle. A **cycle** is the movement from some point (say the far left), to a maximum displacement in the other direction (say the far right) then back to the same point again (the far left). The **period** (T) is simply the time required to complete one cycle. For example, suppose 3 s is required for a bob to move through one complete cycle, to complete the back-and-forth motion from one point, then back to that point. The period of this vibration is 3 s. The number of cycles per second is called the **frequency** (f). For example, a vibrating bob that moves through 1 cycle in 1 s has a frequency of 1 cycle per second. The **length** (L) of the pendulum is measured from the point of suspension to the center of gravity of the bob. Ideally the mass of the bob should be concentrated at a single point. For a small homogeneous sphere, however, the distributed mass is affected by gravity almost as if the mass were all located at the center of the sphere. For all practical purposes then, the length of the pendulum is the *length of the cord plus the radius of the bob* (see figure 6.1).

Procedure

1. Investigate if *weight* influences the period of a pendulum. Use different masses for bobs and adjust each so the length of the string is 100 cm from the pivot point to the center of gravity of the bob. Pull the bob back to make an arc of about 15 degrees. Release and count the number of vibrations (one vibration is one complete back-and-forth movement) for exactly one minute. The arc, length of pendulum, and all other variables must be the same for each run; the only difference should be the mass of the bob. Take three trials for each mass and average the findings. Record all data in Data Table 6.1 on page 70.

2. This time, hold all the variables the same except the *amplitude* of the bob. You can measure the approximate amplitude by using two metersticks as shown in figure 6.2. Count the number of vibrations for exactly one minute for each amplitude tested. Take three trials for each amplitude, average the findings, and record all data in Data Table 6.2 on page 71.

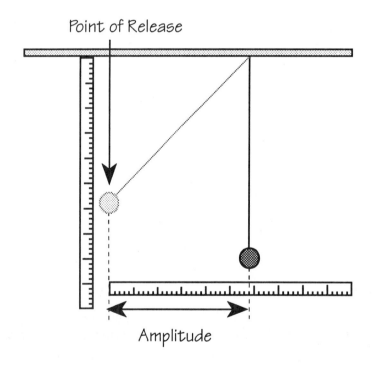

Figure 6.2

3. This time hold all the variables the same except the *length* of the pendulum. Recall that the length of the pendulum is not just the length of the cord. The length is measured from the pivot point of the pendulum to the center of gravity of the bob. Measure the number of vibrations for exactly one

minute. Repeat for three trials for each length, average the findings, and record all data in Data Table 6.3. From the average data, calculate the experimental period for each pendulum length (the time required for one complete vibration) and record it in Data Table 6.3. Use the following equation to calculate the theoretical value of the period for each length and record it on page 72 in Data Table 6.3:

$$T = 2\pi\sqrt{\frac{L}{g}}.$$

4. Use a metal bob to make a pendulum with a 25 cm length. Make three separate measurements of the period of this pendulum by finding the time for the pendulum to vibrate 25 times. Record this data in seconds to the nearest hundredth in Data Table 6.4 on page 73. Multiply the time by 4 to obtain the calculated time of 100 full swings, recording the data in the table. The period in seconds can now be calculated by moving the decimal two places to the left. Use this period and the length of the pendulum in the following equation to calculate the acceleration of the pendulum due to gravity. Then calculate the percentage error and record it in the table.

$$g = \frac{4\pi^2 L}{T^2}$$

Results

1. According to your experimental results, what effect does the weight have on the period of a pendulum?

2. According to your experimental results, what effect does the amplitude have on the period of a pendulum?

3. According to your experimental results, what effect does the length have on the period of a pendulum?

4. Offer a theoretical explanation for all three of the results described for questions 1-3.

5. Was the purpose of this lab accomplished? Why or why not? (Your answer to this question should be reasonable and make sense, showing thoughtful analysis and careful, thorough thinking.)

Going Further

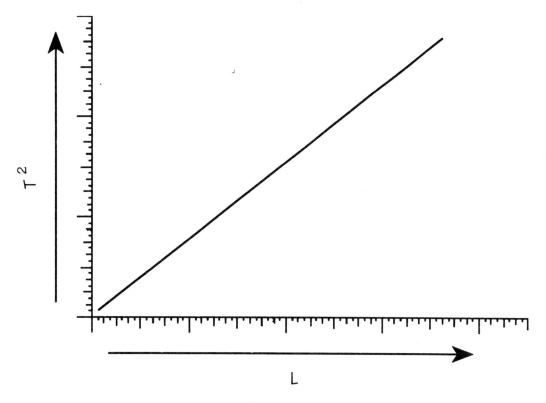

Figure 6.3

If you were to plot T² and L from Data Table 6.4, you would see the relationship illustrated in figure 6.3. The straight-line relation represented in figure 6.3 can be written as

$$T^2 = kL$$

where k is the slope of the line. The true relationship is

$$T = 2\pi\sqrt{\frac{L}{g}}.$$

Therefore, k is equal to

$$\frac{4\pi^2}{g}.$$

Use three data points from procedure step 4 (Data Table 6.4) to plot T^2 versus L, then see if the slope is equal to $4\pi^2/g$. Describe your findings here.

Bob	Cycles per Minute			
	Trial 1	Trial 2	Trial 3	Average
Name: Mass:_____(g)	_____	_____	_____	_____
Name: Mass:_____(g)	_____	_____	_____	_____
Name: Mass:_____(g)	_____	_____	_____	_____
Name: Mass:_____(g)	_____	_____	_____	_____
Name: Mass:_____(g)	_____	_____	_____	_____
Name: Mass:_____(g)	_____	_____	_____	_____

Data Table 6.1   The Effect of Weight on a Pendulum

Data Table 6.2 The Effect of Amplitude on a Pendulum

Approximate Amplitude (cm)	Cycles per Minute			
	Trial 1	Trial 2	Trial 3	Average
_____	_____	_____	_____	_____
_____	_____	_____	_____	_____
_____	_____	_____	_____	_____
_____	_____	_____	_____	_____
_____	_____	_____	_____	_____
_____	_____	_____	_____	_____

Data Table 6.3   The Effect of Length on a Pendulum

Length of Cord plus Radius of Bob (cm)	Cycles per Minute				Experimental Period (s)	Calculated Period (s)
	Trial 1	Trial 2	Trial 3	Average		
_____	_____	_____	_____	_____	_____	_____
_____	_____	_____	_____	_____	_____	_____
_____	_____	_____	_____	_____	_____	_____
_____	_____	_____	_____	_____	_____	_____
_____	_____	_____	_____	_____	_____	_____
_____	_____	_____	_____	_____	_____	_____

Data Table 6.4   The Pendulum and Acceleration Due to Gravity

Trial	Time of 25 full swings (s)	Calculated Time of 100 full swings (s)	Period (T) (s)	Square of Period (T^2) (s)
Example	25.10	100.4	1.004	1.01
1				
2				
3				
Average				

Experimental Value of g: _____

Theoretical Value of g: 980 cm/s^2

Percentage Error: _____

Name_____ Section_____ Date_____

Experiment 7: Projectile Motion

Introduction

There are basically three kinds of motion: (1) the *horizontal*, straight-line motion of objects moving on the surface of the earth; (2) the *vertical* motion of dropped objects that accelerate toward the surface of the earth; and (3) the motion of an object that is projected into the air. The third type of motion, **projectile motion**, could be directly upward as a vertical projection, straight out as a horizontal projection, or at some angle between the vertical and the horizontal. Basic to understanding such compound motion is to understand that (1) gravity always acts on objects, no matter where they are, and (2) the acceleration due to gravity (g) is independent of any motion that an object may have.

As an example of projectile motion, consider figure 7.1. After rolling down the incline AB, the ball moves across a frictionless, horizontal track BC. At C the ball leaves the track to become a projectile. While the ball is still on track BC, and ignoring air resistance, the *speed* of the ball on the track is *constant* because there are *no net forces acting on the ball*. (F = ma; if F = 0, then a = 0.)

After the ball leaves the track, it becomes a projectile. The motion of such a projectile is easier to understand if you split the complete motion into vertical and horizontal parts. After the ball leaves the track, there is an unbalanced force (mg) that accelerates the ball vertically downward. The ball thus has an *increasing downward velocity* the same as that of a dropped ball and is represented by the vertical vector arrows (v_y) in figure 7.2. Ignoring air resistance, there are no forces in the

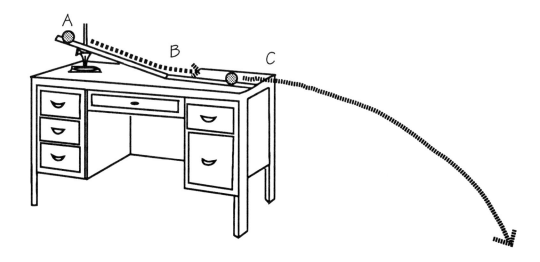

Figure 7.1

horizontal direction so the *horizontal velocity remains the same* as shown by the v_x arrows. The combination of the vertical motion (v_y) and the horizontal motion (v_x) causes the ball to follow a curved path until it hits the floor.

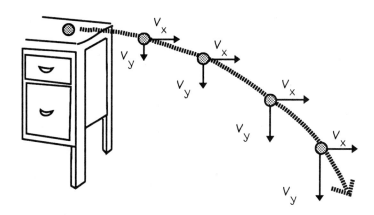

Figure 7.2

The *vertical distance* (Δy) that a falling object moves is proportional to the square of the time that it is falling vertically (t_y). Considering the acceleration due to gravity (g), then

$$\Delta y = \frac{1}{2} g t_y^2$$

The *horizontal distance* (Δx) that a ball moves depends on its horizontal velocity (v_x) when it leaves the horizontal track. Velocity is distance per unit time ($v = d/t$), so

$$v_x = \Delta x / t_x \text{ and } t_x = \Delta x / v_x.$$

Imagine what would happen if the ball had a horizontal velocity only. Without gravity there would be no increasing downward velocity and the ball would move straight out from the table. It would, however, be vertically above where it would have hit the ground (if there were a downward velocity) at the same time. You can see this if you mentally remove the v_y arrows from figure 7.2. Thus the time of fall is determined by t_y, and t_x equals t_y, or $t_x = t_y$.

Since $t_x = t_y$ and $t_x = \Delta x/v_x$, you can substitute $\Delta x/v_x$ for t_y in $\Delta y = \frac{1}{2}gt_y^2$. Then the expression is $\Delta y = \frac{1}{2}g(\Delta x/v_x)^2$ or simply $\Delta y = \frac{1}{2}g(\Delta x^2/v_x^2)$, which now has both the vertical distance (Δy) and the horizontal distance (Δx) in the same relationship. Look at figure 7.3 and think about this relationship.

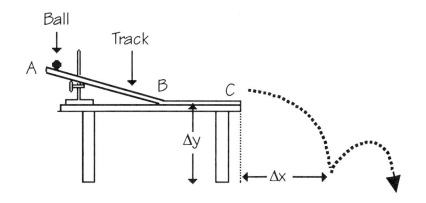

Figure 7.3

Procedure

1. Adjust the ramp so that A, the uppermost position of the ramp, is 10 cm above the surface of the table top. Measure the horizontal part of the ramp labeled BC. Release the ball from A and time how long the ball takes to move across the distance BC. Find the velocity (v_x) of the ball from the relationship $v = d/t$. Make at least three runs and find the average velocity. Record your data in Data Table 7.1.

2. Measure and record Δy. Solve $\Delta y = \frac{1}{2}g(\Delta x^2/v_x^2)$ for Δx, then use the values of Δy and v_x from Data Table 7.1 to find Δx. Record what Δx should be according to your calculation.

3. Place a cup on the floor at the calculated distance Δx from the edge of the table. Roll the ball down the ramp to see if your calculated prediction was correct. You can also use a piece of carbon paper on a sheet of paper for a target.

4. Repeat procedure steps 1 through 3 for ramp heights of 20 cm (Data Table 7.2) and 30 cm (Data Table 7.3), making sure that the position of the stand from the edge of the table is constant.

Results

1. Make a graph with velocity on the *x*-axis and distance on the *y*-axis, and use the average v_x velocity and the average Δx distance from each data table as data points. Calculate the slope, including units, and write the value of the slope here and also somewhere on the graph.

2. Calculate t_y from the square root of $2\Delta y/g$, and discuss why this value should or should not be equal to the slope obtained in question 1 above.

3. Use the results from your calculation in question 2 as the accepted value, and the value of the slope calculated in question 1 as the experimental value, to determine your percentage error. What was the percentage error and probable sources?

4. Discuss how you could improve the precision of this experiment.

5. Was the purpose of this lab accomplished? Why or why not? (Your answer to this question should show thoughtful analysis and careful, thorough thinking.

Data Table 7.1	Projectile Motion from a Ramp: Ramp 10 cm Above Table Top	
Trial	Time (t) (s)	Velocity (v) (cm/s)
1	_____	_____
2	_____	_____
3	_____	_____

Average v_x _____

Vertical Distance Δy _____

Calculated Δx _____

Data Table 7.2	Projectile Motion from a Ramp: Ramp 20 cm Above Table Top	
Trial	Time (t) (s)	Velocity (v) (cm/s)
1	_____	_____
2	_____	_____
3	_____	_____

Average v_x _____

Vertical Distance Δy _____

Calculated Δx _____

Data Table 7.3   Projectile Motion from a Ramp: Ramp 30 cm Above Table Top

Trial	Time (t) (s)	Velocity (v) (cm/s)
1	_____	_____
2	_____	_____
3	_____	_____

Average v_x _____

Vertical Distance Δy _____

Calculated Δx _____

Name_____Section_____Date_____

Experiment 8: Newton's Second Law

Introduction

In this laboratory experiment you will consider **Newton's second law of motion**, which states that an object will accelerate if an unbalanced force acts on it. Upon successful completion of this experiment you should be able to

1. determine the acceleration of a system with a constant mass and a varying force;

2. calculate the mass of the system from experimental data by knowing that the slope of the graph is 1/mass; and

3. calculate the percentage error between the experimental result and the accepted value for a mass.

Newton's second law of motion states the relationship that can be used to predict the change of motion (acceleration) of an object when an unbalanced force is applied. The acceleration is directly proportional to the force and inversely proportional to the mass of the object, or a = F/m.

Recall that the slope of a straight line is a constant obtained from

$$\text{slope (m)} = \frac{\Delta y}{\Delta x}.$$

This equation can be solved for y, with m representing the slope of the straight line. When the origin is zero, the solved equation is as shown in figure 8.1.

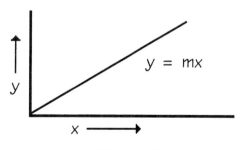

Figure 8.1

From this, you can see that y is directly proportional to x; that is, any increase in x (the manipulated variable) will result in a proportional increase in y (the responding variable).

Figure 8.2 shows three different relationships between three different sets of x and y variables. Each slope has a different proportionality constant (m_1, m_2, and m_3) and each relationship is represented by the equation $y = mx$. In each case the proportionality constant in the equation is equal to the slope of the line.

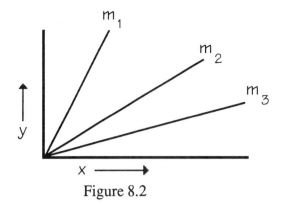

Figure 8.2

In this experiment, you will determine the acceleration of an object by varying the force and keeping the mass constant. From Newton's second law, F = ma, you know that

$$\text{acceleration} = \frac{\text{Force}}{\text{mass}} \quad \text{or} \quad a = \frac{F}{m} \quad \text{or} \quad a = \left(\frac{1}{m}\right)(F).$$

This equation is in the same form as the equation for a straight line (when the origin is equal to zero), or

$$y = mx$$

and

$$a = \left(\frac{1}{m}\right)(F)$$

where y = acceleration, x = force, and the slope (m) is equal to 1/m. Unfortunately, the symbol for slope (m) is the same as the symbol for mass (m). **Don't confuse or try to equate the two symbols.**

Again, recall that the slope of a straight line is determined by

$$\text{slope (m)} = \frac{\Delta y}{\Delta x}.$$

If you plot acceleration (m/sec²) on the y-axis and force (N) on the x-axis, the units for the slope would be

$$\frac{y = m/\sec^2}{x = \text{newtons}} =$$

$$\frac{m/\sec^2}{kg \cdot m/\sec^2} = \frac{m}{\sec^2} \times \frac{\sec^2}{kg \cdot m} = \frac{1}{kg}.$$

The slope therefore is the *reciprocal* of the mass, or 1/kg. As shown in figure 8.3, this is just what you would expect for

$$a = \left(\frac{1}{m}\right)(F).$$

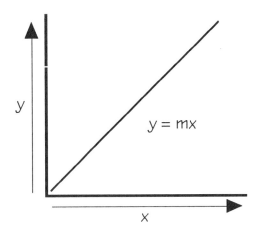 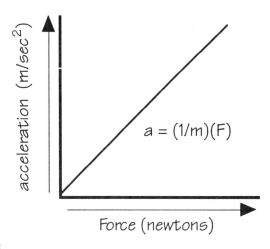

Figure 8.3

Procedure

1. The method of measuring cart or glider velocity may vary with the laboratory setup. You might use a system of a horizontally moving cart attached to weights on a hanger through a string over a pulley as shown in figure 8.4. As the hanger falls, it exerts a horizontal force of F = mg on the cart. Varying the mass on the hanger will thus vary the horizontal force on the cart. Other equipment with other means of measuring velocity might be used in your laboratory, such as a cart with wheels and the use of photogates and computer software. If so, the use of the apparatus will be explained by your instructor.

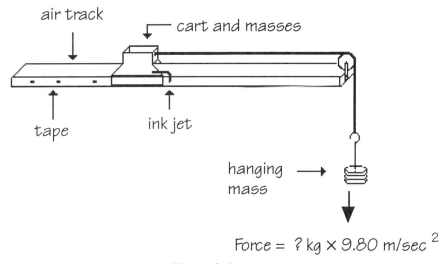

Figure 8.4

2. Tie a nylon string onto the cart. The string should hang about 30 cm down from the pulley when the cart is in the starting position. A weight hanger is attached to the hanger string.

3. Arrange five 50.0 g masses on the cart so they are evenly distributed. You will make six separate runs. On the first run the force on the cart will be the mass of the hanger only times g. For each consecutive run a 50 g mass *is removed from the cart and placed on the hanger.* This

keeps the mass of the system constant but varies the force on the cart. (Note: There are different ways to define a system, and for this lab, the system is defined as the cart, masses, and hanger.) According to the definitions used, the mass of the system is kept constant (while varying the force) by moving masses from the cart and placing them on the hanger. You could add masses to the hanger from outside the system defined, but that is a different experiment.

4. For each of the six runs, calculate the force on the cart from the mass of the hanging masses and hanger in kg times 9.80 m/sec^2. The force will therefore be calculated in *newtons*. Record the force for each run in the appropriate data table.

5. Measure the position of the dots on the tape for each of the six runs. The timer will be set so the dots are 0.05 seconds apart. Carefully measure the distances between the dots and use the change in distance and the time elapsed to calculate the velocity, change in velocity, and acceleration for each run. Record these quantities in the appropriate data tables.

6. Find the average value of the acceleration (a) for each run. Note that the average value must be converted to m/s^2 so the units will be consistent with the units of a newton.

7. Record the average acceleration and the calculated force for each run in the summary data table (8.7). Record the mass of the system (cart, hanger, and masses) in the summary data table.

8. Use the average acceleration and the force values in the summary data table for data points on a graph, with the acceleration in m/s^2 on the *y*-axis and the force in newtons on the *x*-axis. You will have six data points, with each data point corresponding to a run. Calculate the slope of the line (numbers and units) and write it somewhere on the graph.

Results

1. Using the reciprocal of your calculated slope as the experimental value of the mass of the system and the measured mass of the system (cart, hanger, and masses) as found by a balance as the accepted value, calculate the experimental error. Show your calculations here.

2. How do you know if the acceleration was proportional to the force? (Hint: see figure 8.3)

3. How would you know if the acceleration was inversely proportional to the mass?

4. Would the presence of an additional mass in the cart during all the runs increase, decrease, or not affect the slope? How could you use this experimental setup to find the value of the additional mass that was in the cart?

5. Suppose the string breaks as the cart accelerates one-third of the way across the track. What is the acceleration of the cart for the remaining length of track? Explain.

6. Was the purpose of this lab accomplished? Why or why not? (Your answer to this question should be reasonable and make sense, showing thoughtful analysis and careful, thorough thinking.)

Data Table 8.1   Run 1

	d (cm)	Δd (cm)	Δt (s)	v=Δd/Δt (cm/s)	Δv (cm/s)	a = Δv/Δt (cm/s²)
1						
2						
3						
4						
5						

Force = _____(kg)(9.80 m/s²) = _____N

Average acceleration = _____(total) ÷ _____(number) = _____ cm/s²

Average acceleration = _____ m/s²

Data Table 8.2   Run 2

	d (cm)	Δd (cm)	Δt (s)	v=Δd/Δt (cm/s)	Δv (cm/s)	a = Δv/Δt (cm/s²)
1						
2						
3						
4						
5						

Force = _____(kg)(9.80 m/s²) = _____N

Average acceleration = _____(total) ÷ _____(number) = _____ cm/s²

Average acceleration = _____ m/s²

Data Table 8.3		Run 3				
	d (cm)	Δd (cm)	Δt (s)	v=Δd/Δt (cm/s)	Δv (cm/s)	a = Δv/Δt (cm/s²)
1						
2						
3						
4						
5						

Force = _____(kg)(9.80 m/s²) = _____N

Average acceleration = _____(total)÷_____(number) = _____cm/s²

Average acceleration = _____ m/s²

Data Table 8.4		Run 4				
	d (cm)	Δd (cm)	Δt (s)	v=Δd/Δt (cm/s)	Δv (cm/s)	a = Δv/Δt (cm/s²)
1						
2						
3						
4						
5						

Force = _____(kg)(9.80 m/s²) = _____N

Average acceleration = _____(total)÷_____(number) = _____cm/s²

Average acceleration = _____ m/s²

Data Table 8.5 Run 5

	d (cm)	Δd (cm)	Δt (s)	v=Δd/Δt (cm/s)	Δv (cm/s)	a = Δv/Δt (cm/s²)
1						
2						
3						
4						
5						

Force = _____(kg)(9.80 m/s²) = _____N

Average acceleration = _____(total)÷_____(number) = _____cm/s²

Average acceleration = _____ m/s²

Data Table 8.6 Run 6

	d (cm)	Δd (cm)	Δt (s)	v=Δd/Δt (cm/s)	Δv (cm/s)	a = Δv/Δt (cm/s²)
1						
2						
3						
4						
5						

Force = _____(kg)(9.80 m/s²) = _____N

Average acceleration = _____(total)÷_____(number) = _____cm/s²

Average acceleration = _____ m/s²

Data Table 8.7	Summary		
Run	Average Acceleration (m/s^2)	Force (N)	Mass of System (kg)
1			
2			
3			
4			
5			
6			

Name_____Section_____Date_____

Experiment 9: Conservation of Momentum

Introduction

Linear momentum is defined as the product of the mass of an object and its velocity, or

$$p = mv,$$

where p is the symbol for momentum, m is the mass of an object, and v is its velocity. As you can see from this definition, both mass and velocity contribute to the momentum of an object.

Momentum is proportional to the mass and the velocity of an object, which means something must be moving to have momentum (zero velocity times the mass equals zero). If you assume a constant mass of a given object, this means a momentum change of the object is related to a change of velocity. Since a change of velocity (Δv) during a time interval (t) is the definition of acceleration, you could write Newton's second law of motion (F = ma) as

$$F = m\frac{\Delta v}{t}.$$

Multiplying both sides of the equation by t gives

$$Ft = m\Delta v.$$

Since momentum is defined as the product of the mass and velocity (p = mv), this means that a change of momentum (Δp) is equal to mv. Thus $Ft = \Delta p$, and

$$F = \frac{\Delta p}{t}.$$

So a force must be applied for some time interval for a change of momentum to occur. This also means that no net force means that no change of momentum occurs, F = 0 means $\Delta p = 0$. It is a force applied over time that is needed to change the momentum of an object.

When objects are viewed as a system of interacting objects, the total momentum before the interaction is always equal to the total momentum after the interaction. This is expressed as the **law of conservation of momentum**, which states the total momentum of a system of interacting objects is conserved as long as no external forces are involved. For example, the firing of a bullet from a rifle and the recoil of the rifle is a change of momentum in opposite directions. Considering the rifle and the bullet as a system, both the rifle and the bullet have a total momentum of zero with respect to the surface of the earth. When the rifle is fired the exploding gunpowder propels the bullet out with what we could call a forward momentum. At the same time the force from the exploding gunpowder pushes the rifle backward with a momentum opposite to the bullet. The bullet moves forward with a momentum of $(mv)_b$ and the rifle moves backward at the same time. The rifle moves in an opposite direction to the bullet, so its momentum is shown with a minus sign, or $-(mv)_r$. The total momentum of the rifle-bullet system is zero, so

$$\text{Bullet momentum} = -\text{rifle momentum}$$
$$(mv)_b = -(mv)_r$$
$$(mv)_b - (mv)_r = 0.$$

the negative sign simply means a momentum in a direction opposite to the other, and that the momentum of the bullet $(mv)_b$ must equal the momentum of the rifle $-(mv)_r$ in the opposite direction.

The law of conservation of momentum also applies in collisions, such as a speeding car colliding with a stationary car of equal mass, with the coupled cars moving together after the collision. Since momentum is conserved, the total momentum of the system of cars should be the same before and after the collision. Thus

$$\text{Momentum before} = \text{momentum after}$$
$$\text{car 1} + \text{car 2} = \text{coupled cars}$$
$$mv_1 = (m+m) \times \frac{v_1}{2}$$
$$mv_1 = mv_1.$$

(Note that car 2 had zero momentum with a velocity of zero, so there is no mv_2 on the left side of the equation.)

Procedure

In this experiment you will measure the momentum of laboratory carts involved in inelastic collisions. The experiment could be conducted with gliders moving on an air track (figure 9.1), laboratory carts moving on a track cart, or perhaps other types of carts moving across the floor. Your instructor will explain the operation of the equipment you will use.

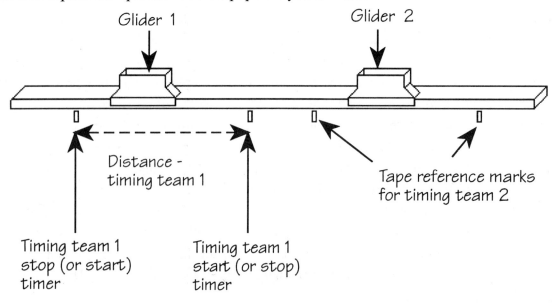

Figure 9.1

The method of measuring cart or glider velocity will vary with the laboratory setup. As shown in figure 9.1, the velocity of a glider or cart before a collision could be measured by measuring the time required to move over a measured distance. Two pieces of tape are used to mark a measured distance *before* the place where a collision will occur. Two other pieces of tape are used to measure the same distance on the track *after* the place where the collision occurs. One team of students, team 1, starts their timers when the front of the cart reaches the first "before collision" reference mark, then stop the timer when the front of the cart reaches the second "before collision" mark. The before collision velocity can be determined from v = d/Δt. A second team of students, team 2 in figure 9.1, does the same procedure for calculating the "after collision" velocity. Other means of measuring velocity that might be used in your laboratory, such as a spark timer or the use of photogates and computer software will be explained by your instructor.

Part A: Equal Mass Collision

1. Find the mass of a cart (or glider) (m_2) and record the mass in Data Table 9.1 on page 104. Place this cart at rest near the center position of the track.

2. Find the mass of another cart (m_1) and record in Data Table 9.1. Place this nearly equal-sized mass near the end of the track.

3. Conduct a trial run by giving m_1 an initial velocity toward m_2 at rest. The cart should have its initial velocity *before* reaching the first reference mark tape (if this method is being used). After the collision, m_1 should be at rest and m_2 should be in motion.

4. Conduct three trials of giving m_1 an initial velocity toward m_2 at rest. Record the velocity before and after each collision and record in Data Table 9.1. Calculate the momentum of m_1 before the collision and the momentum of m_2 after each collision. Record these calculations in Data Table 9.1 for each trial.

5. Since only m_1 was moving before and only m_2 was moving after the collision, these values are the same as the total momentum before and after the collision. Calculate and record in Data Table 9.1 the percent difference in the total momentum before and after the collision for each trial.

Part B: Unequal Mass Collision

1. Add masses to a cart, which we will call m_3, until it is approximately double the mass of m_1. Record the mass of m_1 and m_3 in Data Table 9.2 on page 105. Place m_3 at rest near the center position of the track. Place m_1 near the end of the track.

2. Conduct a trial run by giving m_1 an initial velocity toward m_3 at rest. As before, the m_1 cart should have its initial velocity *before* reaching the first reference mark tape (if this method is being used).

After the collision, m_1 should be moving in an opposite direction from its initial velocity and m_3 should be in motion. This will require the m_1 timing team to measure the velocity both before *and* after the collision.

3. Conduct three trials of giving m_1 an initial velocity toward m_3 at rest. Record the velocity before and after each collision and record in Data Table 9.2. Calculate the momentum of m_1 before the collision and after the collision. Calculate the momentum of m_3 after each collision. Record these calculations in Data Table 9.2 for each trial.

4. Calculate and record in Data Table 9.2 the percent difference in the total momentum before (which is m_1 only) and after the collision for each trial.

Results

1. Describe the possible sources of error in this experiment.

2. Was linear momentum conserved according to the results of this experiment (what did you expect and what did you find)? What is the evidence that supports your response?

3. At the moment that two carts collide, they both experience an instantaneous stop. Is momentum conserved during this stop? Give reasons for your answer.

4. What solid evidence can you provide that the cart was *not* accelerating when it crossed the first reference mark (or photogate)?

5. Was the purpose of this lab accomplished? Why or why not? (Your answer to this question should be reasonable and make sense, showing thoughtful analysis and careful, thorough thinking.)

Data Table 9.1   Momentum of Equal Mass Collision

Trial	m_1 before Collision			m_2 after Collision			Difference
	Mass (g)	Velocity (cm/s)	Momentum (p)	Mass (g)	Velocity (cm/s)	Momentum (p)	Percent (%)
1							
2							
3							

Space for Calculations:

Data Table 9.2   Momentum of Unequal Mass Collision

	m_1 before Collision			Place for Calculations:			
	Mass (g)	Velocity (cm/s)	Momentum (p)				
Trial							
1							
2							
3							

	m_1 after Collision			m_3 after Collision			Total Momentum
	Mass (g)	Velocity (cm/s)	Momentum (p)	Mass (g)	Velocity (cm/s)	Momentum (p)	Percent Difference (%)
Trial							
1							
2							
3							

Name_____Section_____Date_____

Experiment 10: Rotational Equilibrium

Introduction

In addition to straight line, projectile, and motion around a circular path, an object can spin, or rotate about an axis, and this movement is called **rotational motion.** One way to measure how fast something is turning is to count the number of revolutions that occur during a given time. Depending on the nature of what is doing the turning, the frequency of rotation can be expressed in revolutions per second (rps), revolutions per minute (rpm), or revolutions per hour (rph).

What is required to produce rotational motion? If you have ever tried to loosen a stuck bolt with a wrench, you know that where you grasp the wrench, that is, the distance from your hand to the bolt is important. The direction you push is also important, of course, as is the magnitude of the force. These are factors that affect the **torque**, or *tendency of a force to produce a rotation*. As you can see in figure 10.1, applying a torque is the same as applying a force perpendicular to a lever. A lever arm is defined as the perpendicular distance from a center of rotation to an applied force vector. With everything else equal, a force applied farthest from the center of rotation has the longest lever arm and produces the greatest torque (figure10.1). The force applied at right angles (90°) to the lever arm, as opposed to some other angle, also produces the greatest torque. It is often useful to think of a torque as a product obtained by multiplying a force by a lever arm length, or Torque = force × lever arm. In symbols the relationship is

$$\tau = Fr$$

where τ (the Greek letter tau) is the torque, F is the applied force, and r is the perpendicular distance from the center of rotation. Torque units are usually the newton-meter (foot-pound). Torque units are similar to units of energy or work, but they represent completely different concepts.

It is torque, and not force, that should be considered for rotational motion. Two equal but opposite parallel forces on an object would not produce a translation motion because the net force would be zero. It is possible, however, for two equal but opposite parallel forces to result in a net

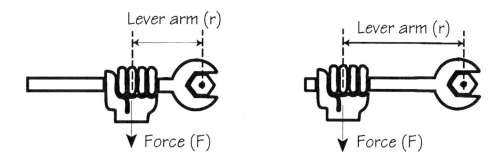

Figure 10.1

force of zero, but produce at the same time a net torque that is greater than zero. For example, a lever with the pivot point in the center will turn if two equal but opposite forces are applied in the same rotational direction. The lever rotates because the torques are greater than zero. There is no translation motion in this situation because the net force is zero.

Newton's laws of motion apply to rotational motion as well as translational motion, but in the case of rotational motion the torque, not the force is considered. Thus for the first law of motion you can see that an object that is not rotating will continue not to rotate as long as the net torque is zero. Likewise, an object that is rotating uniformly will continue to do so at a constant speed as long as no net torque acts to change the motion. In this experiment you will study torques that are needed to keep a body that is not rotating from rotating, that is, torques in rotational equilibrium.

Procedure

A setup for this experiment could be the apparatus show in figure 10.2. It shows a meterstick with a knife-edge clamp on a support stand and masses hanging from movable mass hangers. If this apparatus is not available, a meterstick could be balanced by a loop of string suspended from a support (such as a second meterstick placed on top of two chair backs), with the string tied and held in place with masking tape. The masses could also be suspended by loops of string around the meterstick, which could also be held in place with masking tape. This experiment could be conducted with a wide variety of setups, and the following general procedure can be used. In any case, the first step is to *balance the meterstick*. Record the balance point, which will be called x_0, in Data Table 10.1 on page 113.

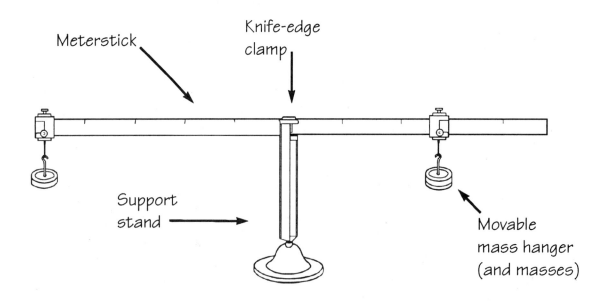

Figure 10.2

Part A: Two Unequal Masses in Equilibrium

1. Hang a 100 g mass (m_1) at a position 50 cm from x_0 on either side of the meterstick. Record the mass of m_1 and its position as x_1 in Data Table 10.1. Note the position of x_1 is the distance from x_0. Also include the mass of any clamp and hanger that may be used with m_1.

2. Hang a 200 g mass (m_2) on the opposite side of the meterstick from m_1. Adjust the position of m_2 until static equilibrium is found. Record the mass of m_2 and its position as x_2 in Data Table 10.1. Again, note the position of x_2 is the distance from x_0 and also be sure to include the mass of any clamp and hanger that may be used with m_2.

3. Calculate the torque τ_1 around x_0 from the force of gravity on the 100 g mass, $\tau_1 = m_1 g x_1$, where g is 9.80 m/s², the average acceleration of gravity. Show your calculations here and record τ_1 in Data Table 10.1.

4. Calculate the torque τ_2 around x_0 from the force of gravity on the 200 g mass, $\tau_2 = m_2 g x_2$, where g is 9.80 m/s². Show your calculations here and record τ_2 in Data Table 10.1.

5. Compare the clockwise and counterclockwise torques on a sketch in Data Table 10.1 and record the percent difference. Record your calculations and comments, if any, here:

Part B: Three Unequal Masses in Equilibrium

1. Hang a 50 g mass (m_1) and a 100 g mass (m_2) at two different positions *on the same side* of the meterstick and record the position of x_1 from x_0 and the position of x_2 from x_0 in Data Table 10.2 on page 114. Also include the mass of any clamp and hanger that may be used with m_1 and m_2.

2. Hang a 200 g mass (m_3) on the opposite side of the meterstick from m_1 and m_2. Adjust the position of m_3 until static equilibrium is found. Record the mass of m_3 and its position as x_3 in Data Table 10.2. Again, be sure to include the mass of any clamp and hanger that may be used with m_3.

3. Calculate the torque τ_1 around x_0. Show your calculations here and record τ_1 in Data Table 10.2.

4. Calculate the torque τ_2 around x_0. Show your calculations here and record τ_2 in Data Table 10.2.

5. Calculate the torque τ_3 around x_0. Show your calculations here and record τ_3 in Data Table 10.2.

6. Compare the clockwise and counterclockwise torques on a sketch in Data Table 10.2 and record the percent difference between ($\tau_1 + \tau_2$) and τ_3. Record your calculations and comments, if any, here:

Part C: Unknown Mass in Equilibrium

1. Hang an unknown mass (m_1) on one side of the meterstick, but near x_0, and record the position of x_1 from x_0 in Data Table 10.3 on page 115.

2. Hang a 100 g mass (m_2) on the opposite side of the meterstick from m_1. Adjust the position of m_2 until static equilibrium is found. Record the mass of m_2 and its position as x_2 in Data Table 10.3. Include the mass of any clamp and hanger that may be used with m_2.

3. Calculate and compare the torque τ_2 from the mass m_2 and distance x_2 to the torque τ_1 from the unknown mass and the known distance x_1. Use the equation $\tau_1 = \tau_2$ to solve for the unknown mass m_1. Show your calculations here and record your findings in Data Table 10.3.

4. Use a laboratory balance to find the mass of m_1, recording the measurement in Data Table 10.3.

5. Compare the value of the mass of m_1 as determined from the clockwise and counterclockwise torques to the value of the mass of m_1 as determined by the laboratory balance. Calculate the percent difference between these two methods in Data Table 10.3. Record your calculations and comments here:

Results

1. What is a torque? Why are torques, not just forces considered for rotational equilibrium?

2. Explain why masses are moved back and forth, from notch to notch, along a scale on a laboratory balance.

3. Compare the measurement errors that might result from determining an unknown mass as in Part C of this experiment with the measurement errors that might result from using a laboratory balance to determine the mass.

4. Was the purpose of this lab accomplished? Why or why not? (Your answer to this question should be reasonable and make sense, showing thoughtful analysis and careful, thorough thinking.)

Data Table 10.1	Two Unequal Masses in Equilibrium
Balance point (x_0)	
Mass (m_1)	
Distance of x_1 and x_0	
Torque τ_1	
Mass (m_2)	
Distance of x_2 and x_0	
Torque τ_2	
Percent difference in τ_1 and τ_2	
Sketch showing labeled clockwise and counterclockwise torques:	

Data Table 10.2	Three Unequal Masses in Equilibrium
Mass (m_1)	
Distance of x_1 and x_0	
Torque τ_1	
Mass (m_2)	
Distance of x_2 and x_0	
Torque τ_2	
Mass (m_3)	
Distance of x_3 and x_0	
Torque τ_3	
Percent difference in ($\tau_1 + \tau_2$) and τ_3	
Sketch showing labeled clockwise and counterclockwise torques:	

Data Table 10.3	Unknown Mass in Equilibrium
Distance of x_1 and x_0	
Torque τ_1	
Mass (m_2)	
Distance of x_2 and x_0	
Torque τ_2	
Calculated (m_1)	
Measured (m_1) on laboratory balance	
Percent difference	
Sketch showing labeled clockwise and counterclockwise torques:	

Name_____Section_____Date_____

Experiment 11: Centripetal Force

Introduction

    This experiment is concerned with the force necessary to keep an object moving in a constant circular path. According to Newton's first law of motion there *must be* forces acting on an object moving in a circular path since it does not move off in a straight line. The second law of motion (F = ma) also indicates forces since an unbalanced force is required to change the motion of an object. An object moving in a circular path is continuously being accelerated since it is continuously changing direction. This means that there is a continuous unbalanced force acting on the object that pulls it out of a straight-line path. The force that pulls an object out of a straight-line path and into a circular path is called a **centripetal force.**

    The magnitude of the centripetal force required to keep an object in a circular path depends on the inertia (or mass) and the acceleration of the object, as you know from the second law (F = ma). The acceleration of an object moving in uniform circular motion is $a = v^2/r$, so the magnitude of the centripetal force of an object with a mass (m) that is moving with a velocity (v) in a circular orbit of radius (r) can be found from

$$F = \frac{mv^2}{r}.$$

The distance (circumference) around a circle is $2\pi r$. The velocity of an object moving in a circular path can be found from $v = d/t$, or $v = 2\pi r/T$, where $2\pi r$ is the distance around one complete circle and T is the period (time) required to make one revolution. Substituting for v,

$$F = \frac{m\left(\frac{2\pi r}{T}\right)^2}{r}$$

or

$$F = \frac{\frac{m 4\pi^2 r^2}{T^2}}{r},$$

$$F = \frac{4\pi^2 r^2 m}{T^2} \times \frac{1}{r},$$

$$F = \frac{4\pi^2 r\, m}{T^2}.$$

This is the relationship between the centripetal force (Fc), the mass (m) of the object in circular motion, the radius (r) of the circle, and the time (T) required for one complete revolution.

Procedure

1. The equipment setup for this experiment consists of weights (washers) attached to a string, and a rubber stopper that swings in a horizontal circle (fig. 11.1). You will swing the stopper in a circle and adjust the speed so that the stopper does not have a tendency to move in or out, thus balancing the centripetal force (F_c) on the stopper with the balancing force (F_b), or mg, exerted by the washers on the string.

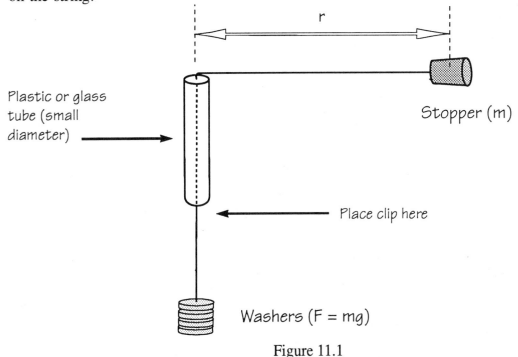

Figure 11.1

2. Place some washers on the string and practice rotating the stopper by placing a finger next to the string, then moving your hand in a circular motion. You are trying to move the stopper with a consistent, balancing motion, just enough so the stopper does not move in or out. *Keep the stopper moving in a fairly horizontal circle, without the washers moving up or down.* An alligator (or paper) clip placed on the string just below the tube will help you maintain a consistent motion by providing a point of reference as well as helping with length measurements. Be careful of the moving stopper so it does not hit you in the head.

3. After you have learned to move the stopper with a constant motion in a horizontal plane, you are ready to take measurements. The distance from the string at the top of the tube to the *center* of the stopper is the radius (r) of the circle of rotation. The mass (m) of the stopper is determined with a balance. The balancing force (F_b) of the washers is determined from the mass of the washers times g ($F_b = mg$). The period (T) is determined by measuring the time of a number of revolutions, then dividing the total time by the number of revolutions to obtain the time for one revolution. For example, 20 revolutions in 10 seconds would mean that $^{10}/_{20}$, or 0.5 seconds, is required for one revolution. This data is best obtained by one person acting as a counter speaking aloud while another person acts as a timer.

4. Make four or five trials by rotating the stopper with a different number of washers on the string each time, adding or removing two washers (about 20 g) for each trial. For each trial, record in Data Table 11.1 the mass of the washers, the radius of the circle, and the average time for a single revolution.

Data Table 11.1	Centripetal Force Relationships				
Trial	Mass of Washers (m) (kg)	Balancing Force (F_b) (N)	Radius (r) (m)	Time (t) (s)	Centripetal Force (F_c) (N)
1					
2					
3					
4					
5					

Mass of stopper _____ kg

5. Calculate and record the balancing force (F_b) for each trial from the mass of washers times g (9.80 m/s^2), or $F_b = mg$.
6. Calculate and record the centripetal force (F_c) for each trial from

$$F = \frac{4\pi^2 r\, m}{T^2}.$$

Considering the balancing force (F_b) as the accepted value, and the calculated centripetal force (F_c) as the experimental value, calculate your percentage error for each trial of this experiment. Analyze the percentage errors and other variables to identify some trends, if any.

Trial 1 :

Trial 2 :

Trial 3 :

Trial 4 :

Trial 5 :

Results

1. Did the balancing force (F_b) equal the centripetal force (F_c)? Do you consider them equal or not equal? Why or why not?

2. Analyze the errors that could be made in all the measured quantities. What was probably the greatest source of error and why? Discuss how these errors could be avoided and how the experiment in general could be improved.

3. Discuss any trends that were noted in your analysis of percentage error for the different trials. Analyze the meaning of any observed trends or discuss the meaning of the lack of any trends.

4. Was the purpose of this lab accomplished? Why or why not? (Your answer to this question should be reasonable and make sense, showing thoughtful analysis and careful, thorough thinking.)

Name_____Section_____Date_____

Experiment 12: Archimedes' Principle

Introduction

Archimedes' principle states that an object floating or submerged in a liquid is buoyed up by a force equal to the weight of the liquid displaced by the object. The purpose of this experiment is to study this principle as it applies to floating and submerged objects and its application determination of specific gravity.

The buoyant force on an object immersed in a liquid can be determined by weighing an object in air and then in water. The apparent loss of weight in water, or $W_w - W_a$, where W_a is the weight in air and W_w is the weight in water, is the buoyant force of the water. The weight of the water displaced by an object can be measured by using an overflow can and catch bucket (figure 12.1). The relationship between the buoyant force on an immersed object can then be compared to the weight of the water displaced.

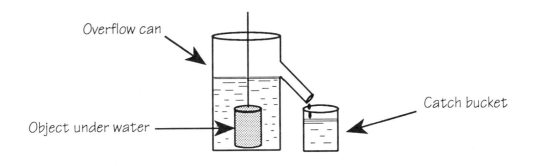

Figure 12.1

The specific gravity of an object is defined as the ratio of the density of the object to the density of water at the same temperature. According to Archimedes' principle, the apparent loss of weight of an object immersed in a liquid is equal to the weight of the liquid displaced. The specific gravity of an object more dense than water is easily determined by weighing the object in air, then weighing it suspended in water. The loss of weight in water is equal to the weight of the water displaced, that is, the weight of an equal volume of water. The loss of weight in water is $W_w - W_a$, where W_a is the weight in air and W_w is the weight in water, so the specific gravity (S) is equal to

$$S = \frac{W_a}{W_a - W_w}$$

123

Procedure

1. Weigh the object in air and record the weight in Data Table 12.1.

2. Fill an overflow can with water while holding a finger over the spout hole. Place the can on the table, with the spout hole over a catch bucket. Remove your finger from the spout hole, catching the excess water. When the water in the can is level with the spout, discard the excess water from the catch bucket and dry it with paper towels. Record the weight of the dry catch bucket in Data Table 12.1. Place the dry catch bucket beneath the spout hole again.

3. Weigh the object in water by attaching it by a fine thread to a spring scale or laboratory balance, then lowering it into the overflow can. Immerse the object completely, catching the overflow water in the catch bucket as shown in figure 12.1. Record the weight in Data Table 12.1.

4. Complete the calculations in Data Table 12.1.

5. Repeat procedure steps 1-4, this time with a floating object such as a block of wood. Record the measurements and calculations in Data Table 12.2.

Results

1. Compare the weight of the *submerged object,* the buoyant force on the object, and the weight of the water displaced by the object:

2. Compare the weight of the *floating object,* the buoyant force on the object, and the weight of the water displaced by the object:

3. What generalization would describe your comparisons in questions 1 and 2?

4. What is the specific gravity of the submerged object according to the measurements in Data Table 12.1, as compared to the known value (if available)?

5. Was the purpose of this lab accomplished? Why or why not? (Your answer to this question should show thoughtful analysis and careful, thorough thinking.)

Going Further

Explain why a ball of clay sinks, but when shaped into a boat, it floats. Experiment with a ball of clay and container of water to find which shape of clay will hold the most "cargo."

Data Table 12.1 Buoyant Force on Submerged Object	
1. Weight of object in air	_____
2. Apparent weight of object in water	_____
3. Buoyant force on object (row 1 minus row 2)	_____
4. Weight of empty catch bucket	_____
5. Weight of catch bucket plus overflow water	_____
6. Weight of displaced water (row 4 minus row 5)	_____

Data Table 12.2 Buoyant Force on Floating Object	
1. Weight of object in air	
2. Apparent weight of object in water	
3. Buoyant force on object (row 1 minus row 2)	
4. Weight of empty catch bucket	
5. Weight of catch bucket plus overflow water	
6. Weight of displaced water (row 4 minus row 5)	

Name_____Section_____Date_____

Experiment 13: Hooke's Law

Introduction

Considering what happens to a solid when it is squeezed or stretched leads to the conclusion that molecules have strong forces of interaction. Some materials are deformed, or "bent out of shape" by squeezing or stretching forces, but return to their original shape when the squeezing or stretching force is removed. Some materials do this better than others and the materials that do it best have the property of elasticity. Elastic materials return to their original shape after being deformed by some external force—as long as the force was not too great. If you press gently on the exterior metal side of a car door, you can see the metal return to its original shape when you stop pressing—if you didn't press too hard. Elastic materials such as the metal of your car door have intermolecular forces that pull the molecules back to their original positions. This pulls the solid as a whole back to its original shape when the applied force is removed. Such responses to compression or stretching forces lead to the assumption that molecules repel one another if pushed closer together and attract one another if pulled farther apart.

There is a limit to how much external force a given material can experience and still have the intermolecular forces pull or push the material back to its original shape. The elastic limit is the maximum force per unit area that a material can be subjected to before becoming permanently deformed. At the elastic limit the intermolecular forces are overcome and the molecules slide past one another, permanently altering the shape of the object. If enough force is applied the material will break or fracture. At the fracture point the intermolecular forces are overcome to such an extent that the molecules are completely separated. It is the property of elasticity that causes a rubber ball to bounce. The force of striking the floor deforms the ball when it is dropped. As the ball regains its original shape, it pushes itself away from the floor causing the ball to rise above the floor, or bounce. A clay ball does not bounce because it exhibits plastic deformation. Plastic is the opposite of elastic. A material does not recover from a plastic deformation and stays in the shape into which it was deformed. The materials called plastics were so named because they exhibit this property when warm. Both elastic and plastic deformations take place before the fracture point is reached.

When a coiled spring is stretched by a weight, it is deformed but will return to its original shape and length. If the spring is stretched too far, however, the elastic limit may be exceeded and the spring will be permanently stretched out of shape. Before the elastic limit is reached, a spring will stretch by a length that is proportional to the weight pulling on it. For example, if a one pound weight stretches the spring one inch, a two pound weight will stretch it two inches, and a three pound weight will stretch it three inches. You could say that the deformation is directly proportional to the applied force within the elastic limit. The directly proportional relationship between the stretch, or change in length ΔL, and the applied force F is called **Hooke's Law**. In symbols,

$$F = -\Delta k L$$

where k is a proportionality constant depending on the elastic material. Note that k is positive; the sign is negative because the internal force exerted by the elastic materials is in a direction opposite to the applied force, that is, the force exerted by the spring and ΔL are in opposite directions. The larger k is, the greater force (F) needed to cause a given ΔL.

The proportionality constant, k, may be measured for any object by applying a known force, F, to cause a measurable deformation, ΔL. The restoring force will be equal in magnitude to the applied force, and k can thus be determined by using

$$k = \frac{F}{\Delta L}.$$

As you can see, k will have units of N/m. Solved for F, and with ΔL expressed as the change of length from the original length ($L_f - L_i$), the equation becomes

$$F = k(L_f - L_i)$$
or
$$F = (kL_f - kL_i).$$

Note this equation is in the same form as the equation for a straight line, which has a general form of

$$y = mx + b,$$

where m is the slope and b is the y-intercept.

Procedure

1. Hang a spring, small end up, to a rigid support. Place a 50 g weight hanger on the lower end. Initially, enough weight must be placed on the hanger so that no two sections of the spring touch each other and all the kinks are out of the spring. The length of the straightened spring and the attached weight will be used as a starting point for all subsequent length and weight measurements. Measure this length from the top of the spring to the bottom of the weight hanger and record the length in Data Table 13.1 on page 133.

2. Add 0.1 kg to the weight hanger. Measure and record the new length to the bottom of the weight hanger.

3. Take a total of ten more sets of measurements covering a total distance of at least 0.1 m. Record all measurements in the data table.

Results

1. Make a graph of each data point, with the total distance on the *x*-axis and and the total weight on the *y*-axis. Calculate the slope, including units, and write the value of the slope here and also somewhere on the graph.

2. Use the data to calculate an average *k* from mg/ΔL. Discuss why this value should or should not be equal to the slope obtained in question 1.

3. Use the results from your calculation in question 2 and the value of the slope calculated in question 1 to determine a percent difference. What were the percentage difference and probable sources of experimental error?

4. Discuss how you could improve the precision of this experiment.

5. Was the purpose of this lab accomplished? Why or why not? (Your answer to this question should show thoughtful analysis and careful, thorough thinking.)

Going Further

Using the same spring and setup of this experiment, adjust the total mass on the spring to 500 g. Pull the spring down 5 cm and release it. Measure its period by timing 5 to 10 oscillations or cycles. Divide the total time by the number of cycles to find the average period. The equation for the period of a spring is

$$T = 2\pi\sqrt{\frac{m}{k}}$$

where $\pi = 3.1416$, m = mass, and k = the spring constant. Use this equation to determine the spring constant k, then compare it to the value found from the spring elongation graph in the experiment.

Table 13.1   Hooke's Law

Mass of straightened spring and hanger masses	_____ kg
Length of spring and weight hanger	_____ m
Length of spring with ____0.1____ kg added	_____ m
Length of spring with _____ kg added	_____ m
Length of spring with _____ kg added	_____ m
Length of spring with _____ kg added	_____ m
Length of spring with _____ kg added	_____ m
Length of spring with _____ kg added	_____ m
Length of spring with _____ kg added	_____ m
Length of spring with _____ kg added	_____ m
Length of spring with _____ kg added	_____ m
Length of spring with _____ kg added	_____ m
Length of spring with _____ kg added	_____ m
Average k _____	

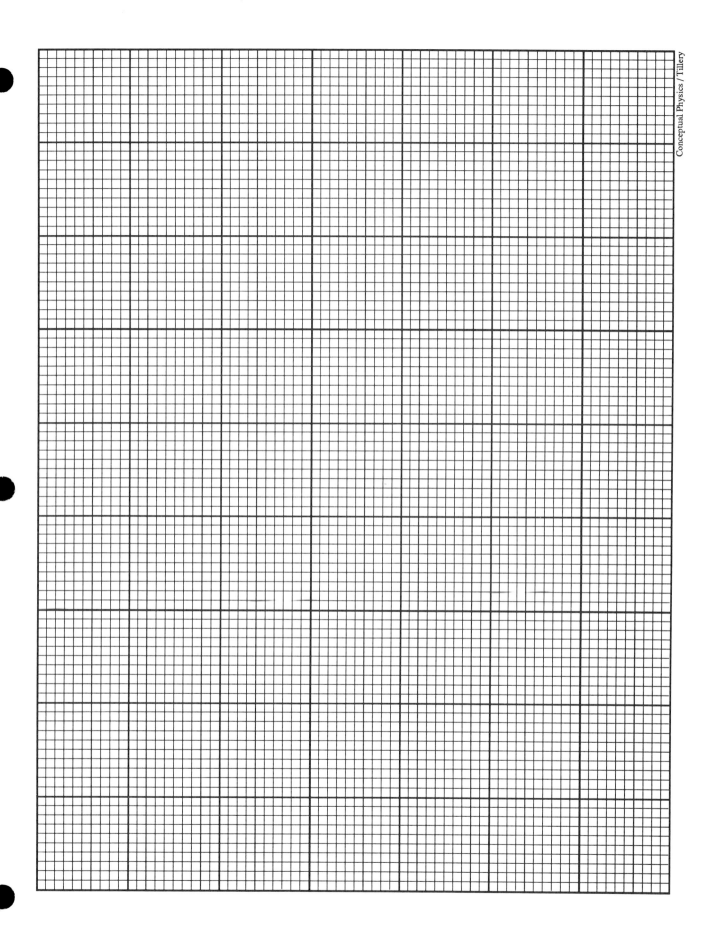

Name_____ Section_____ Date_____

Experiment 14: Young's Modulus

Introduction

Within the elastic limit, Hooke's law (see experiment 13) tells us that a force distorts a body by some amount proportional to the force doing the stretching. Recalling the molecular structure of a solid, you can imagine that greater and greater forces are trying to pull the molecules of the solid farther and farther apart (the stretch), which causes greater and greater forces between the molecules trying to pull everything back together again. You know that the forces must match since the system is in equilibrium, that is, the weight exerting the force is not moving up or down. Thus the downward force exerted by a weight on a spring is matched by an upward force exerted by internal forces between the molecules of the spring.

The internal molecular forces that tend to resist changes in shape or volume are called *stress*. **Stress** is defined as the internal forces within a body that result from the application of an external force. It can be expressed as the ratio of the applied external force to the area over which the force acts, or

$$\text{Stress} = \frac{\text{Force}}{\text{Area}}.$$

In symbols the equation is

$$\text{Stress} = \frac{F}{A}.$$

When a force acts perpendicular to a surface, the force per unit area is called *pressure*. Pressure on a body does result in stress. However, pressure is not the whole story since the direction as well as the magnitude of a force can vary and this results in different kinds of stresses. The simplest kinds of stress are *compression*, *tension*, and *shear*.

Local stress always results in some amount of compression, pulling, or twisting of matter, and these changes are called *strain*. **Strain** is the deformation caused by stress. For solids under tension or compression stress, the resulting strain can be expressed as a ratio of the resulting change of length to the original, unstressed length, or

$$\text{Strain} = \frac{\text{change of length}}{\text{original length}}.$$

In symbols this is

$$\text{Strain} = \frac{\Delta L}{L}$$

where ΔL is the change in the length due to the stress and L is the original unstressed length. Note that ΔL will be negative from compression stress and positive from tension stress. Since this compression or tension strain is defined as a ratio of a change of length to the original length, it has no units. Note there are other kinds of strain that result from the other kinds of stress. When you restate Hooke's law in terms of an applied stress and the resulting strain, you understand that *strain is proportional to the stress*.

Hooke's law provides the means of measuring the elasticity of different materials. Recall that Hooke's law describes the amount of change in the shape of a solid object as being directly proportional to the force applied to it, or

$$F = -kL$$

where F is the force exerted, L is the amount of elongation or compression, and k is a constant that is determined by the material being used (the minus means the spring force is opposite the displacement). Solving for k, we have

$$k = \frac{F}{L}.$$

The force producing a deformation can be defined in terms of stress. The resulting deformation can also be defined in terms of strain, so we see that Hooke's law can also be expressed as

$$k = \frac{F}{L} = \frac{\text{stress}}{\text{strain}}.$$

You could carry this equation one step further since we already have an equation that defines stress and another that defines strain. Thus,

$$k = \frac{\text{stress}}{\text{strain}} = \frac{\dfrac{\text{force applied}}{\text{area of cross section}}}{\dfrac{\text{elongation}}{\text{original length}}}$$

or in symbols,

$$k = \frac{\text{stress}}{\text{strain}} = \frac{F/A}{\Delta L/L}$$

The ratio of an imposed stress and the resulting strain will have the same units as stress since the strain units cancel. In physics, "modulus" means a measure of the extent that a substance possesses some property. The **stretch modulus** describes how well an object retains its length when a stretching (or compressing) stress is applied. It is also a measure of stiffness. A small stretch modulus value means that a material is easily stretched since less stress is needed to achieve a given strain. A large modulus value means that a material is not easily stretched, or is stiff since more stress is needed to achieve the strain. The stretch modulus is sometimes referred to as *Young's modulus* after one of the early investigators of elasticity. In this experiment you will compute Young's modulus—a ratio of the stress to the strain—for a steel wire.

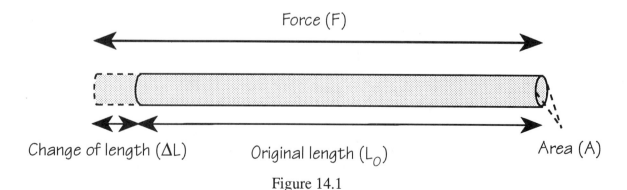

Figure 14.1

Procedure

This experiment requires a Young's modulus apparatus, which is illustrated in figure 14.2. The wire to be tested is fastened at the top of the apparatus and a mass hanger is fastened to the wire at the bottom. Masses are added and removed from the mass hanger to vary the force that produces the stress. The resulting strain is small, so some means of magnifying the changes is needed. One way of magnifying the change is to use an "optical lever," a mirror device that attaches to the wire and becomes tilted one way or the other as the wire length changes. A low-power laser is usually used to bounce a beam of light off the mirror to a vertical paper attached to a wall about two meters away. The reflected beam moves up and down on the paper with changes in the length of the wire. The change of length of the wire, ΔL, can be calculated from

$$\Delta L = \frac{d \Delta y}{2D}$$

where d is the distance from the moving foot of the optical lever to a line between the front feet, D is the distance between the mirror and the paper on the wall, and Δy is the distance the spot of light moves on the wall paper when a mass is added or removed from the hanger. Your instructor will demonstrate this method, or a different method of measuring the strain if an alternate is to be used.

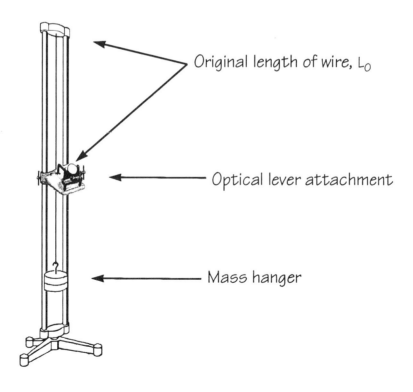

Figure 14.2

1. Add a 1 kg mass to the hanger on a steel wire. Measure the length of wire between the fastener at the top of the apparatus down to the top of the fastener on the optical lever attachment. Record this length as the initial length of the wire (L_0) in Data Table 14.1.

2. Measure the wire diameter with a micrometer caliper, avoiding any kinks in the wire. Make three measurements, average the results, and record the average diameter in Data Table 14.1. Use the average diameter to calculate the cross-sectional area of the wire in square meters, and record in Data Table 14.1.

3. Place the optical lever mount on a piece of paper and carefully make a sketch of the three feet. Draw a line between the front feet, then use a vernier caliper to measure the perpendicular distance between the line and the rear foot. Record this distance as d in Data Table 14.1. Return the mount to the apparatus, with the two front feet in the groove and the rear foot in the wire holder. Adjust the mirror to a vertical orientation.

4. Set the laser so a horizontal beam strikes the mirror, then reflects horizontally over an exactly two meter distance to the center of a white strip of paper mounted vertically to a wall or screen. Record this distance as D in Data Table 14.1. Mark the position of the light beam on the paper. **CAUTION: Do not look into the laser beam.** (Your instructor might use a different procedure, depending on the available equipment.)

5. Increase the load on the steel wire by adding 1 kg masses, marking the position of the light beam on the paper each time. Identify each mark to show the number added, increasing the load. Do not hurry the measurements since the wire needs a little time to adjust to load changes. For the steel wire (only), add a total of 10 1-kg masses to the hanger.

6. Decrease the load on the steel wire by removing 1 kg masses, again marking the position of the light beam on the paper. Again identify each mark, this time identified to show a decreasing load.

7. When only one 1-km mass remains, turn the laser off and measure the length of wire between the fastener at the top of the apparatus down to the top of the fastener on the optical lever attachment. Record this length as the final length of the wire (L_0). Average the initial and final L_0 and record in Data Table 14.1.

8. With the laser remaining off, use a vernier caliper to measure the distance between the marks on the paper screen, Δy, as the load was increased and decreased. Record in Data Table 14.2 all the measurements for increasing and decreasing load increments. Calculate and record an average Δy for each load (increasing and decreasing).

9. Use $\Delta L = d\Delta y/2D$ to calculate a change of length (ΔL) for each added mass. Use $F = mg$, where g is 9.80 m/s^2, to calculate the stress (F/A) from each added mass. Calculate the strain ($\Delta L/L_0$) for each load. Record each change of length, stress, and strain in Data Table 14.2.

10. Make a graph of stress (F/A) versus strain ($\Delta L/L_0$) and draw a best fit straight-line curve. Calculate the slope and record it in Data Table 14.1. Compare this value to the accepted value for steel. Calculate the percent error and record it in Data Table 14.1.

Results

1. What are several names used to identify the property you calculated by finding the slope? What does this property tell you about a material?

2. Are the elasticity and the strength of a material the same property? Explain.

3. Describe the meaning of stress and strain in your own words.

4. Was the purpose of this lab accomplished? Why or why not? (Your answer to this question should be reasonable and make sense, showing thoughtful analysis and careful, thorough thinking.)

Data Table 14.1	Elasticity of Steel Wire
Initial L_0	
Final L_0	
Average L_0	
Diameter of wire	
Trial 1	
Trial 2	
Trial 3	
Average	
Area of wire	
Distance d from optical lever	
Distance D from mirror	
Young's modulus from slope	
Young's modulus from reference	
Percent error	

Data Table 14.2 Young's Modulus for Steel Wire							
Load (kg)	Δy Increasing Load	Δy Decreasing Load	Average Δy	Change of Length ΔL	Strain $\frac{\Delta L}{L_0}$	Stress $\frac{F}{A}$
2						
3						
4						
5						
6						
7						
8						
9						
10						

Name_____ Section_____ Date_____

Experiment 15: Friction

Introduction

Friction is the resisting force that opposes the sliding or rolling motion of one body over another. The force needed to overcome friction depends on the nature of the materials in contact, their smoothness, and on the normal force. Imagine a large crate on a horizontal floor. You want to move it by pushing it to the right with a horizontal force, F (figure 15.1). If F is small, the frictional force (f) acting to the left will keep the crate from moving (F = f) and the crate will remain in equilibrium. Since the crate is not moving, this frictional force is called the **force of static friction** (f_s). If the magnitude of F is gradually increased, F will eventually exceed f_s and the crate will move to the right. Once you start the crate moving, you will find that it takes the force F´ to keep it moving at a constant velocity. The crate is in equilibrium so there must be no unbalanced force on it and the force tending to stop the motion must be equal to and opposite the force F´. The retarding frictional force for an object in motion is called the **force of kinetic friction** (f_k). Since the crate moves with a constant speed, then you know that F´ = f_k. If F´ were to increase as to become unbalanced, the crate would accelerate to the right. If F´ were to decrease, f_k would now be larger and the crate would accelerate to rest. Both f_s and f_k are proportional to the normal force and both depend on the nature of the surfaces in contact. This relationship can be summarized by the following equation,

$$F = \mu N$$

where F is the applied force opposing friction and μ (the Greek letter mu) is called the **coefficient of friction**. As you can see, μ is the fraction of the normal force that it takes to make surfaces slide over

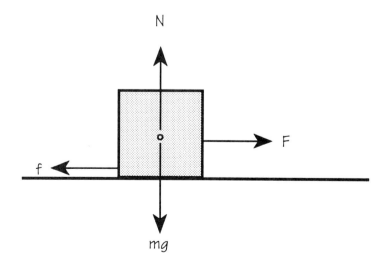

Figure 15.1

each other. In this particular example the normal force is the weight but this is not always the case. Normal means perpendicular, *not vertical*. So, we can write the equilibrium condition as

$$\Sigma F = 0$$

and so

$$0 = F - |\mu N|$$

or

$$F = |\mu N|$$

The absolute value signs are around the term with the N because the N is perpendicular to the direction of F so one cannot legitimately set F = N times a constant. Two vectors cannot be equal if they are not parallel, but this symbol is often not written. The coefficient of friction that opposes the direction of motion is called the coefficient of sliding friction or the **coefficient of kinetic friction**. It is the one used when the two surfaces are moving relative to one another and often has the symbol μ_k. The coefficient of kinetic friction could also be expressed as a ratio of the force of kinetic friction to the normal force produced by two surfaces pressing together, or

$$\mu_k = \frac{F}{N}$$

where F is the force of kinetic friction directed parallel to the surfaces and opposite to the direction of motion, N is the normal force, and μ_k is the coefficient of kinetic friction. In cases where the two surfaces are *not moving* relative to one another we use the **coefficient of static friction**, μ_s. This coefficient can be expressed as a ratio of the force of static friction (F_s) to the normal force necessary to start movement, or

$$\mu_s = \frac{F_s}{N}$$

In this investigation, you will use these relationships to determine the coefficient of static friction and the coefficient of kinetic friction between two wood surfaces.

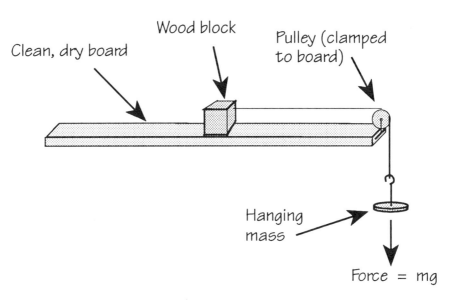

Figure 15.2

Procedure

1. Use a balance to find the mass of the wood block. Record the mass in Data Table 15.1 on page 151, then calculate and record the weight of the block.

2. Place a clean, dry board on the laboratory table with its pulley over the edge of the table (fig. 15.2). Place the block on the board with the largest area in contact with the board. A light cord is attached to the block and a weight hanger, then run over the pulley. Add small masses to the hanger until the force created is just sufficient to keep the block moving slowly with a constant speed after it has been started with a gentle push. Record this force (mg) in Data Table 15.1.

3. Repeat procedure step 2, this time placing increasing masses on top of the block and recording the force needed to keep the block moving slowly with a constant speed when it has been started with a gentle push. Record in Data Table 15.1 the mass on the block along with the force needed to keep the block moving at a constant speed.

4. Place a 500 gram mass (or masses) on top of the block. Gradually increase the mass of the hanger until the block moves, without a push, with a uniform speed. If the block accelerates, start over and use slightly less mass on the hanger. Do this three times, and average the results. Record the average force needed (mg) in Data Table 15.2 on page 152.

5. Use the information in Data Table 15.1 to plot the values of the force of friction versus the values of the normal force (see figure 15.1). Calculate the slope to obtain the coefficient of kinetic friction μ_k for wood on wood. Write the value here and somewhere on the graph.

6. Use the information in Data Table 15.2 to find the coefficient of static friction μ_s for wood on wood. Use the equation

$$\mu_s = \frac{F_s}{N}$$

and show your calculations here.

Results

1. Which was greater, the coefficient of static friction or the coefficient of kinetic friction? Is this the result you were expecting? Explain.

2. Why was it necessary for the block to move with a constant velocity in all procedures?

3. How consistent were the friction effects observed in procedure step 4? Why would this be the case?

4. Was the purpose of this lab accomplished? Why or why not? (Your answer to this question should be reasonable and make sense, showing thoughtful analysis and careful, thorough thinking.)

Data Table 15.1	Coefficient of Kinetic Friction			
Mass Placed on Block (kg)	Mass of Block (kg)	Total Weight of Block and Masses (N)	Total Normal Force (N)	Force Needed to Move Block Uniformly (N)
_____	_____	_____	_____	_____
_____	_____	_____	_____	_____
_____	_____	_____	_____	_____
_____	_____	_____	_____	_____
_____	_____	_____	_____	_____
_____	_____	_____	_____	_____

Coefficient of kinetic friction (from graph) _____

Data Table 15.2	Coefficient of Static Friction			
	Mass Placed on Block	Weight Placed on Block	Total Normal Force	Force Required to Move Block	Coefficient of Static Friction
Trial 1	_____	_____	_____	_____	_____
Trial 2	_____	_____	_____	_____	_____
Trial 3	_____	_____	_____	_____	_____
Average	_____	_____	_____	_____	_____

Going Further

How would changing the area of contact between the block and the board affect (a) the μ_k and (b) the μ_s? Make your prediction below, then experiment with one of the smaller surfaces of the block to test your prediction.

Conceptual Physics / Tillery

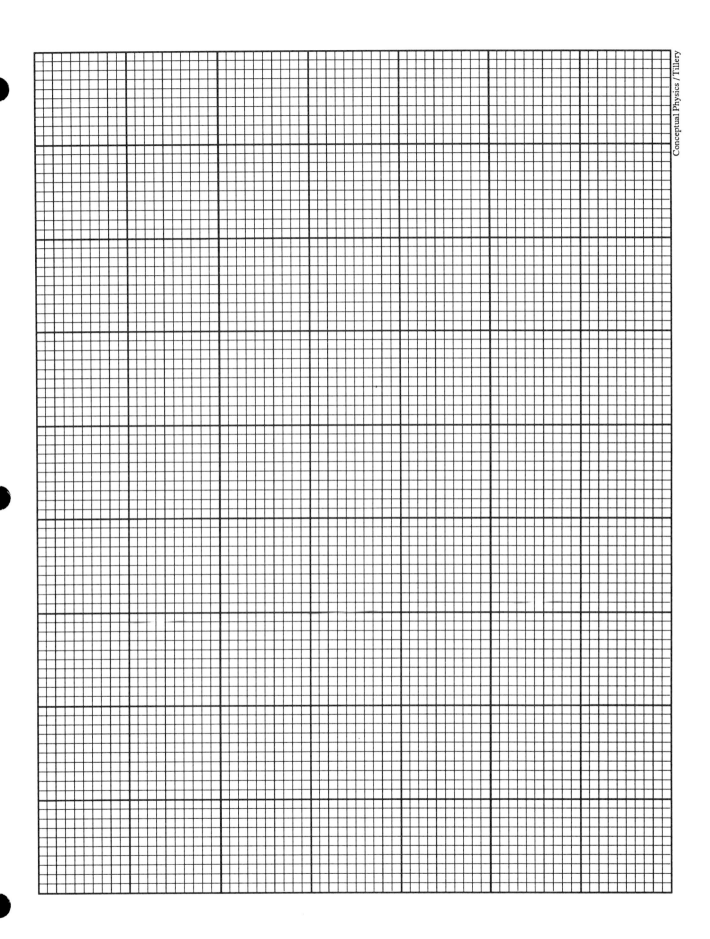

Name_____Section_____Date_____

Experiment 16: Work and Power

Introduction

The word *work* represents a concept that has a special meaning in science that is somewhat different from your everyday concept of the term. In science, the concept of work is concerned with the application of a force to an object and the distance the object moves as a result of the force. **Work** (*W*) is defined as the magnitude of the applied force (F) multiplied by the distance (d) through which the force acts, $W = Fd$.

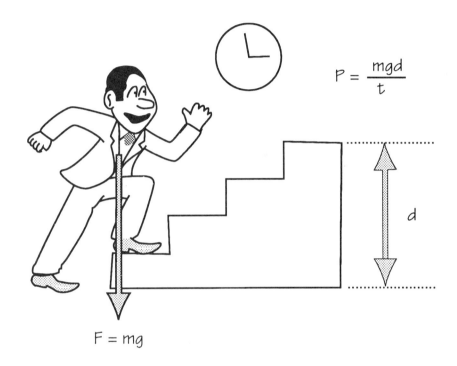

Figure 16.1

You are doing work when you walk up a stairway since you are lifting yourself through a distance (fig. 16.1). You are lifting your weight (the force exerted) the vertical height of the stairs (distance through which the force is exerted). Running up the stairs rather that walking is more tiring because you use up your energy at a greater rate when running. The rate at which energy is transformed or the rate at which work is done is called power. **Power** (*P*) is defined as work (*W*) per unit of time (*t*),

$$P = \frac{W}{t}.$$

When the steam engine was first invented there was a need to describe the rate at which the engine could do work. Since people at that time were familiar with using horses to do their work, the steam engines were compared to horses. James Watt, who designed a workable steam engine, defined **horsepower** (hp) as a power rating of 550 ft·lb/s. In SI units, power is measured in joules per second, called the **watt** (W). It takes 746 W to equal 1 hp, and 1 kW is equal to about 1⅓ hp.

Procedure

1. Teams of two volunteers will measure the work done, the rate at which work is done, and the horsepower rating as they move up a stairwell. Person A will measure and record the data for person B. Person B will measure and record the data for person A. An ordinary bathroom scale can be used to measure each person's weight. Record the weight in pounds (lb) in Data Table 16.1. This weight is the force (F) needed by each person to lift himself or herself up the stairs.

2. The vertical height of the stairs can be found by measuring the height of one step, then multiplying by the number of steps in the stairs. Record this distance (d) in feet (ft) in Data Table 16.1.

3. Measure and record the time required for each person to *walk normally* up the flight of stairs. Record the time in seconds (s) in Data Table 16.1.

4. Measure and record the time required for each person to *run* up the flight of stairs as fast as can be safely accomplished. Record the time in seconds (s) in Data Table 16.1.

5. Calculate the work accomplished, power level developed, and horsepower of each person while walking and while running up the flight of steps. Be sure to include the correct units when recording the results in Data Table 16.1.

Results

1. Explain why there is a difference in the horsepower developed in walking and running up the flight of stairs.

2. Is there some limit to the height of the flight of stairs used and the horsepower developed? Explain.

3. Could the horsepower developed by a slower-moving student ever be greater than the horsepower developed by a faster-moving student? Explain.

4. Describe an experiment that you could do to measure the horsepower you could develop for a long period of time rather than for a short burst up a stairwell.

5. Was the purpose of this lab accomplished? Why or why not? (Your answer to this question should show thoughtful analysis and careful, thorough thinking.)

Data Table 16.1   Work and Power Data and Calculations

	Volunteer A		Volunteer B	
	Walking	Running	Walking	Running
Weight (F) (lb)				
Vertical height (d) of steps (ft)				
Time required (t) to *walk* the flight of steps (s)				
Time required (t) to *run* the flight of steps (s)				
Work done $W = Fd$				
Power $P = W/t$				
Horsepower developed $P \div 550$ ft·lb/s				

Name_____Section_____Date_____

Experiment 17: Thermometer Fixed Points

Introduction

This experiment is concerned with the fixed reference points on the Fahrenheit (F) and Celsius (C) thermometer scales. Two easily reproducible temperatures are used for the fixed reference points and the same points are used to define both scales. The fixed points are the temperature of melting ice and the temperature of boiling water under normal atmospheric pressure. The differences in the two scales are (1) the numbers assigned to the fixed points, and (2) the number of divisions, called **degrees**, between the two points. On the Fahrenheit scale, the value of 32 is assigned to the lower fixed point and the value of 212 is assigned to the upper fixed point, with 180 divisions between these two points. On the Celsius scale, the value of 0 is assigned to the lower fixed point and the value of 100 is assigned to the upper fixed point, with 100 divisions between these two points. In this laboratory investigation you will compare observed thermometer readings with the actual true fixed points.

Variations in atmospheric pressure have a negligible effect on the melting point of ice but have a significant effect on the boiling point of water. Water boils at a higher temperature when the atmospheric pressure is greater than normal, and at a lower temperature when the atmospheric pressure is less than normal. Normal atmospheric pressure, also called **standard barometric pressure**, is defined as the atmospheric pressure that will support a 760 mm column of mercury. An atmospheric pressure change that increases the height of the column of mercury will increase the boiling point by 0.037° C (0.067° F) for each 1.0 mm of additional height. Likewise, an atmospheric pressure change that decreases the height of the column will decrease the boiling point by 0.037° C (0.067° F) for each 1.0 mm of decreased height. Thus you should add 0.037° C for each 1.0 mm of a laboratory barometer reading above 760 mm and subtract 0.037° C for each 1.0 mm below the normal pressure of 760 mm. This calculation will give you the actual boiling point of water under current atmospheric pressure conditions. Any difference between this value and the observed thermometer reading is an error in the thermometer.

Procedure

1. First, verify accuracy of the lower fixed point of the thermometer. Fill a beaker with cracked ice as shown in figure 17.1. After water begins forming from melting ice, place the bulb end of the thermometer well into the ice, but leave the lower fixed point on the scale uncovered so you can still read it. Gently stir for five minutes and then until you observe no downward movement of the mercury. When you are confident that the mercury has reached its lowest point, carefully read the temperature. The last digit of your reading should be an estimate of the distance between the smallest marked divisions on the scale. Record this observed temperature of the melting point in

Data Table 17.1. Use 0° C as the accepted value and calculate and record the measurement error, if any.

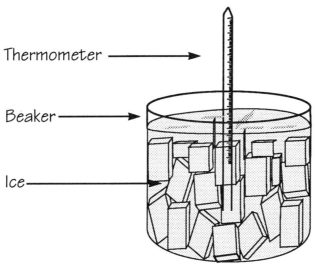

Figure 17.1

2. Now verify the accuracy of the upper fixed point of the thermometer. Set up the steam generator as illustrated in figure 17.2. If you need to insert the thermometer in the stopper, be sure to moisten both with soapy water first. Then hold the stopper with a cloth around your hand and *gently* move the thermometer with a twisting motion. The water in the steam generator should be adjusted so the water level is about 1 cm below the thermometer bulb. When the water begins to boil vigorously, observe the mercury level until you are confident that it has reached its highest point. Again, the last digit of your reading should be an estimate of the distance between the smallest

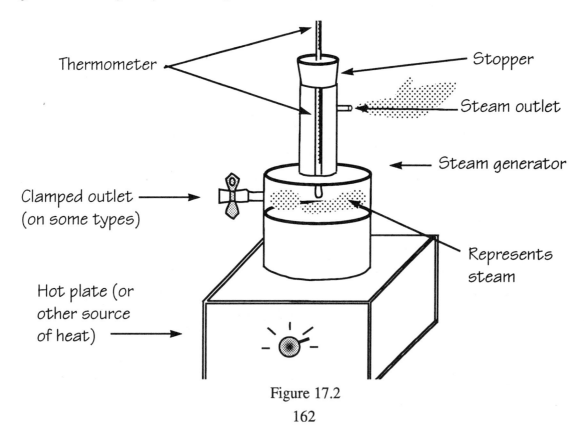

Figure 17.2

marked divisions on the thermometer scale. Record this observed temperature of the boiling point in Data Table 17.1.

3. Determine the accepted value for the boiling point by recording in mm the barometric pressure, then calculating the deviation above or below 100° C. Record this accepted boiling point in Data Table 17.1, then calculate and record the measurement error, if any.

4. Repeat the entire procedure for a second trial, recording all data in Data Table 17.1.

Results

1. Did the temperature change while the ice was melting? Offer an explanation for this observation.

2. Describe how changes in the atmospheric pressure affect the boiling point of water. Offer an explanation for this relationship.

3. Account for any differences observed in the melting point and boiling point readings.

4. How would the differences determined in this investigation influence an experiment concerning temperature if the errors were not considered?

5. Was the purpose of this lab accomplished? Why or why not? (Your answer to this question should show thoughtful analysis and careful, thorough thinking.)

Going Further

Using data from your *best* trial, make a graph by plotting the Celsius temperature scale on the x-axis and the Fahrenheit temperature scale on the y-axis. Calculate the slope of the straight line and write it here and on the graph somewhere, then answer the following questions:

1. What is the value of the slope? What is the meaning of the slope?

2. What is the value of the y-intercept?

3. The slope-intercept form for the equation of a line is $y = mx + b$, where y is the variable on the y-axis (in this case, °F), x is the variable on the x-axis (in this case, °C), m is the slope of the line, and b is the y-intercept. Use this information to write the equation of the Celsius-Fahrenheit temperature graph. What is the meaning of this equation?

Data Table 17.1   Thermometer Readings and Actual Fixed Points

	Trial 1	Trial 2
Observed Melting Point		
Measurement Error—Melting Ice		
Observed Boiling Point		
Barometric Pressure		
Deviation from Normal (+ or −)		
Accepted Boiling Point		
Measurement Error—Boiling Water		

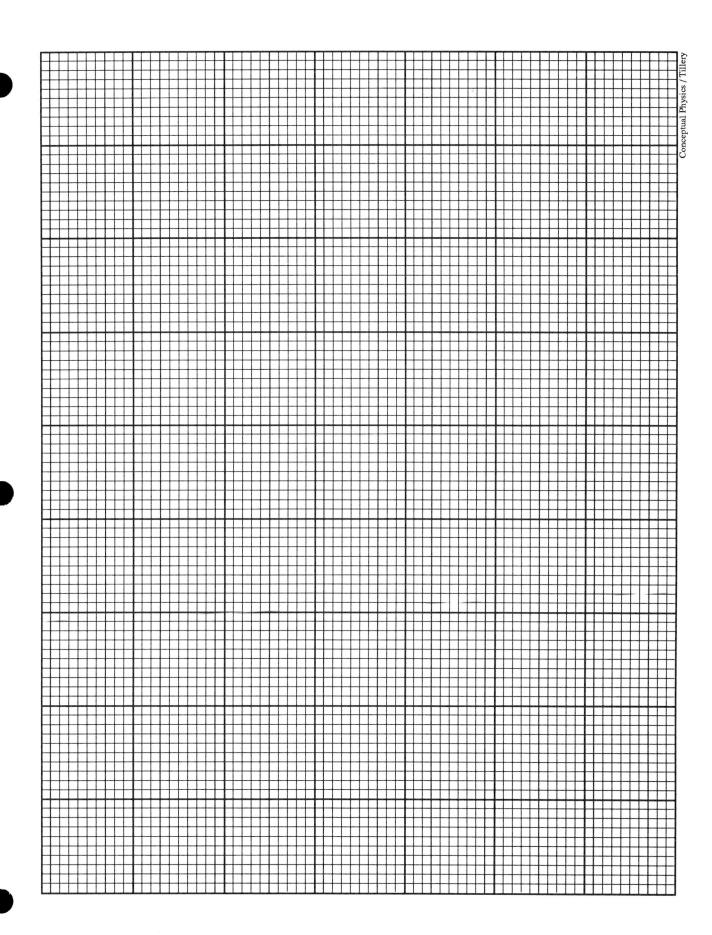

Name_____ Section_____ Date_____

Experiment 18: Absolute Zero

Introduction

If you place an inflated balloon in a freezer, you will find that the balloon becomes much smaller as it cools. Warming the balloon back to its original temperature results in an increased volume, which will be the original volume if no air leaked during the cooling and warming process. The volume of air in the balloon depends on the temperature of the air and increases or decreases in the volume are proportional to increases or decreases in the temperature. The change in volume is proportional to the change in temperature, or V ∝ T.

Since the volume of a gas decreases with decreases in temperature, how much of a temperature decrease would be needed to decrease the volume of the gas to zero? A liter of gas loses about 1/273 of its volume for each degree it is cooled from 0° C. So, you might project a zero volume occurring at a temperature of –273° C. This will not happen, of course, since the gas will liquefy and then perhaps solidify as the temperature is lowered. There are many other relationships, however, that point to –273° C as a special temperature. It is the coldest temperature possible, the complete lack of heat, and it is called **absolute zero**. The Kelvin scale has the same degree interval size as the Celsius scale but begins at absolute zero. The Kelvin scale is more than a Celsius scale with the zero point moved down by 273 degrees. The Kelvin scale is absolute, not relative to some arbitrary fixed points as are the Celsius and Fahrenheit scales. Thus, zero on the Kelvin scale *does* mean zero, and all the numbers above zero have meaning in relation to one another. This relationship does not exist on the relative number scales.

Thus, if you double the absolute temperature of a gas, its volume doubles as V ∝ T. If the temperature remains constant, the volume of an enclosed gas is inversely proportional to the pressure, or

$$V \propto \frac{1}{P}.$$

Combining these two relationships,

$$V \propto \frac{T}{P} \quad \text{or} \quad PV \propto T.$$

By inserting a proportionality constant k, you can write the relationship as an equation that is known as the **ideal-gas law**,

$$PV = kT \quad \text{where T is in Kelvins.}$$

You will use the ideal-gas law in this experiment. You will observe a small glass tube that is closed at one end and has a movable bead of mercury near the other end. The air pressure inside the tube increases or decreases with changes in temperature. The atmospheric pressure is constant, so the mercury bead moves back and forth with the changing pressure of the air trapped inside. Because the small glass tube is of uniform diameter, the length of the air column should be proportional to PV, which should, in turn, be proportional to the temperature.

Procedure

1. The small glass tube with the mercury bead is placed inside a larger tube so hot and cold fluids can flow evenly and completely around the small tube, uniformly changing the temperature of the air inside the tube (figure 18.1).

2. Your laboratory instructor will adjust the position of the mercury bead so it is about two-thirds of the total length from the closed end.

3. You will measure the length of the column of air inside the small tube by measuring from the closed end of the tube to the inside edge of the mercury bead. Measure the bead at a right angle to the length of the tube in all trials. It is also important to measure the length of the column of air from the side of the bead that is closest to the closed end of the tube. The mercury will expand and contract with changes in temperature, as well as change positions with the column of air.

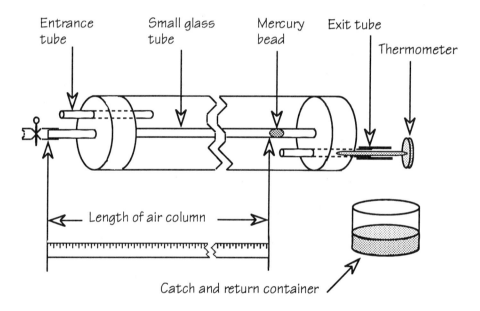

Figure 18.1

4. Measure the length of the air column inside the small tube at room temperature. Record this length and the room temperature in Data Table 18.1.

5. Adjust the large tube so the exit end is slightly higher than the entrance end (figure 18.2). The tube should slope slightly upward, allowing air bubbles to leave the tube as you run water through it. Run ice water through the large tube, catching and returning the water with a catch container, then pouring it back over the funnel full of ice. Have a second container ready to catch water while you are pouring water back through the funnel. Continue recycling the water through the ice until the temperature at the exit is constant. Record this temperature and the length of the air column in Data Table 18.1.

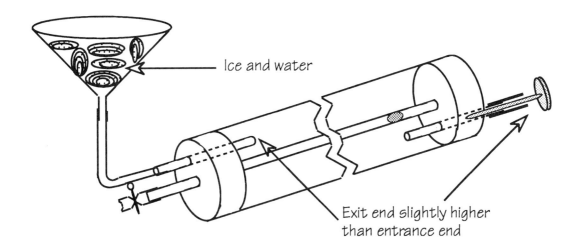

Figure 18.2

6. With the tube still sloping slightly upward to allow the escape of air bubbles, run hot water through the system until the temperature at the exit is constant. Record this temperature and the length of the air column in Data Table 18.1.

7. Adjust the large tube so it slopes slightly downward (figure 18.3). Run steam through the system until the temperature at the exit is constant. The downward slope will permit water that condenses from the steam to escape. Record the temperature and the length of the air column in Data Table 18.1.

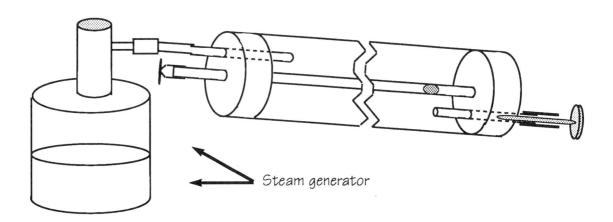

Figure 18.3

Data Table 18.1	Volume of Air and Temperature Relationships	
Fluid	Temperature (°C)	Length of air column (cm)
Room Air		
Ice Water		
Hot Water		
Steam		

Results

1. Analysis by extrapolation: Make a graph with the *x*-axis running from –350° C to +120° C. Plot the data points and make a straight best-fit line in the region of the data points. Using a ruler, continue with a dotted line until it crosses the *x*-axis. The dotted line will cross the *x*-axis at the projected, or extrapolated, Celsius temperature of absolute temperature.

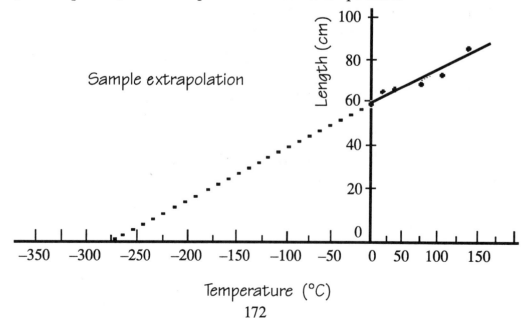

Sample extrapolation

2. Analysis from slope: Make a second graph with the x-axis running from 0° C to 100° C. Plot the data points and draw a best-fit line. Determine the slope of the line. Determine the y-intercept. Use the slope-intercept equation of a straight line, $y = mx + b$, to find x when $y = 0$. Since y is the length of the air column and x is the temperature, then

$$0 = (\text{slope})(\text{temperature}) + b$$

or

$$\text{temperature}\,(0\text{ K}) = \frac{-b}{\text{slope}}.$$

3. Find the experimental error of analysis by extrapolation, using –273.15° C as the accepted value of absolute zero.

4. Find the experimental error of analysis by equation of slope, using –273.15° C as the accepted value of absolute zero.

5. Was the purpose of this lab accomplished? Why or why not? (Your answer to this question should show thoughtful analysis and careful, thorough thinking.)

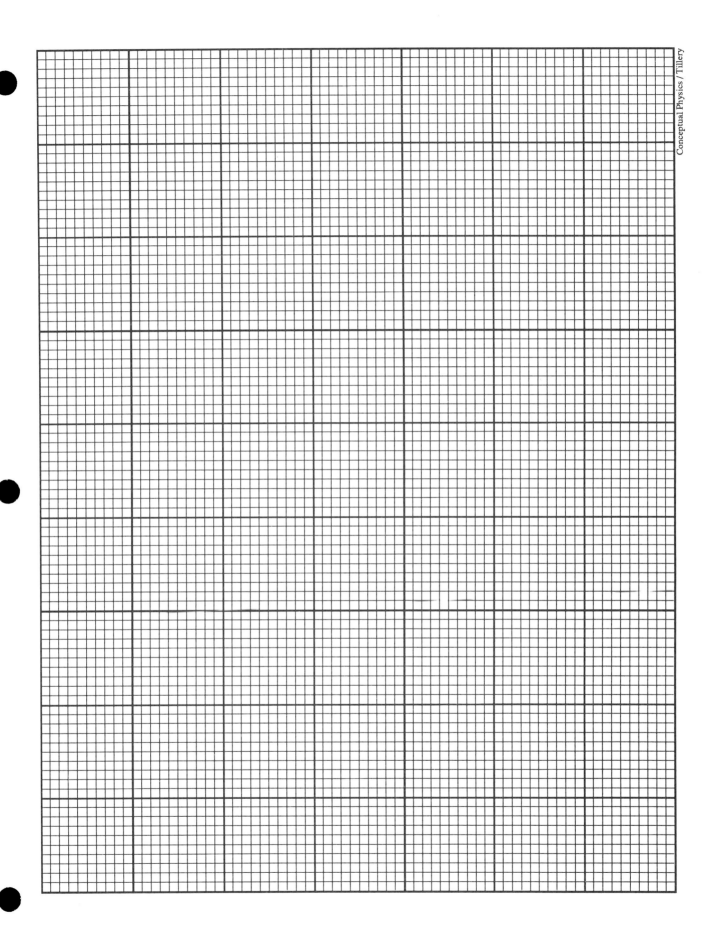

Name_____Section_____Date_____

Experiment 19: Coefficient of Linear Expansion

Introduction

Most substances expand when heated through a normal temperature range. The change in length (ΔL) when a solid is heated is found to be proportional to the initial length (L_0) and to the change in temperature (ΔT), or

$$\Delta L = \alpha L_0 \Delta T.$$

The constant α is called the average **coefficient of linear expansion**, which is defined as the change in length of each unit of length when the temperature is changed one degree. The value of the coefficient of linear expansion depends on the material of which the solid is made. To determine the value of α for different materials, which is the purpose of this investigation, the equation can be solved for the coefficient α:

$$\alpha = \frac{\Delta L}{L_0 \Delta T}$$

or

$$\alpha = \frac{L - L_0}{L_0 (T - T_0)}$$

where L_0 is the initial length, L is the new length that results from an increase in temperature, T_0 is the initial temperature, and T is the final temperature in degrees Celsius. Note that the average coefficient of linear expansion has units of $1/C°$.

There are several different types of laboratory devices that can be used to determine the coefficient of linear expansion for a metal rod. All types employ some means of magnifying the small expansion that occurs for the range of temperatures used. Most types of devices use room temperature as the initial temperature and the temperature of steam from a small laboratory boiler, to which a metal rod or tube is heated, as the final temperature. In one common type of device steam flows around a rod, which expands and pushes against the short arm of a lever, causing the longer arm to swing through a greater distance across a calibrated scale. Another common type of device sends steam from a laboratory boiler through a metal tube. The tube is fixed at one end, and as the heat expands the tube, the other end moves across a pin that rotates a pointer over a dial. Your laboratory instructor will describe the exact procedure for the type of laboratory device used.

Procedure

1. Measure the length of the copper rod (tube) to 0.1 mm. Record this initial length and the room temperature in Data Table 19.1.

2. Set up the laboratory apparatus according to the instructions of your laboratory instructor. Heat water in the boiler, allowing steam to pass through the tube or apparatus until measurement shows no further expansion. If the boiler has a thermometer, measure and record the temperature to the nearest tenth of a degree. Otherwise, note that the temperature of steam is 100° C at normal pressure. Measure and record the change of length of the copper rod (tube).

3. Repeat the above procedures for an aluminum rod (tube).

4. Repeat the procedures for an iron rod (tube).

Results

1. Calculate the coefficients of linear expansion of copper, aluminum, and iron from your data. Record your work here:

2. Compare your findings with the coefficients of linear expansion listed in a reference table. Calculate the percentage error and record your work here:

3. When the metal rod is heated, does it expand in length only? Explain.

4. What are the possible sources of error in this investigation?

5. Was the purpose of this lab accomplished? Why or why not? (Your answer to this question should show thoughtful analysis and careful, thorough thinking.)

Going Further

1. The Golden Gate Bridge in San Francisco is 1,280 m long. Assume the bridge is all made of steel and the temperature increases from 0° C to 40° C. Find the coefficient of linear expansion for steel in a reference book, then compute the total expansion of the bridge for the conditions given.

2. As a metal pipe is heated, does the radius of the interior opening increase or decrease? Explain.

Data Table 19.1 Coefficient of Linear Expansion of Metals			
	Copper	Aluminum	Iron
Room Temperature			
Length of Rod at Room Temperature			
Change of Temperature			
Change of Length of Rod			
Coefficient of Linear Expansion			
Percent Error			

Name_____Section_____Date_____

Experiment 20: Specific Heat

Introduction

Heating is a result of energy transfer, and a quantity of heat can be measured just as any other quantity of energy. The metric unit for measuring energy or heat is the **joule**. However, the separate historical development of the concepts of motion and energy and the concepts of heat resulted in separate units. Some of these units are based on temperature differences.

The metric unit of heat is called the **calorie** (cal), a leftover term from the old caloric theory of heat. A calorie is defined as the amount of energy (or heat) needed to increase the temperature of one gram of water one degree Celsius. A kilocalorie (kcal) is the amount of energy (or heat) needed to increase the temperature of one kilogram of water one degree Celsius. The relationship between joules and calories is called the **mechanical equivalence of heat,** and the relationship is

$$4.184 \text{ J} = 1 \text{ cal}$$
or
$$4184 \text{ J} = 1 \text{ kcal}.$$

There are three variables that influence the energy transfer that takes place during heating: (1) the temperature change, (2) the mass of the substance being heated, and (3) the nature of the material being heated. The relationships among these variables are:

1. The quantity of heat (Q) needed to increase the temperature of a substance from an initial temperature of T_i to a final temperature of T_f is proportional to $T_f - T_i$, or $Q \propto \Delta T$.
2. The quantity of heat (Q) absorbed or given off during a certain ΔT is also proportional to the mass (m) of the substance being heated or cooled, or $Q \propto m$.
3. Differences in the nature of materials result in different quantities of heat (Q) being required to heat equal masses of different substances through the same temperature range.

The **specific heat** (c) is the amount of energy (or heat) needed to increase the temperature of one gram of a substance one degree Celsius. The property of specific heat describes the amount of heat required to heat a certain mass through a certain temperature change, so the units for specific heat are cal/g°C or kcal/kg°C. Note that the k's in the second set of units cancel, so the numerical value for both is the same — for example, the specific heat of aluminum is 0.217 cal/g°C, or 0.217 kcal/kg°C. Some examples of specific heats in these units are:

Aluminum 0.217	Iron 0.113
Copper 0.093	Silver 0.056
Lead 0.031	Nickel 0.106

When the units of all three sets of relationships are the same units used to measure Q, then all the relationships can be combined in equation form,

$$Q = mc\Delta T.$$

This relationship can be used for problems of heating or cooling. A negative result means that energy is leaving a material; that is, the material is cooling. When two materials of different temperatures are involved in heat transfer and are perfectly insulated from their surroundings, the heat lost by one will equal the heat gained by the other,

$$\text{Heat lost}_{\text{(by warm substance)}} = \text{Heat gained}_{\text{(by cool substance)}}$$
or
$$Q_{\text{lost}} = Q_{\text{gained}}$$
or
$$(mc\Delta T)_{\text{lost}} = (mc\Delta T)_{\text{gained}}.$$

Calorimetry consists of using the concept of conservation of energy and applying it to a mixture of materials initially at different temperatures that come to a common temperature. In other words,

(heat lost by sample) = (heat gained by water).

The sample is heated, then placed in water in a calorimeter cup where it loses heat. The water is initially cool, gaining heat when the warmer sample is added. (The role of a Styrofoam calorimeter cup in the heat transfer process can be ignored since two Styrofoam cups have negligible heat gain [$\Delta T \approx 0$] and very little mass.) In symbols,

$$m_s c_s \Delta T_s = m_w c_w \Delta T_w$$

where m_s is the mass of the sample, c_s the specific heat of the sample, and ΔT_s is the temperature change for the sample. The same symbols with a subscript w are used for the mass, specific heat, and temperature change of the water. Solving for the specific heat of the sample gives

$$c_s = \frac{m_w c_w \Delta T_w}{m_s \Delta T_s}.$$

Procedure

1. You are going to determine the specific heat of three samples of different metals by using calorimetry. You will run two trials on each sample, making *very careful* temperature and mass measurements. Do the calculations before you leave the lab. If you have made a mistake you will still have time to repeat the measurements if you know this before you leave.

2. Be sure you have sufficient water to cover at least the bottom two-thirds of a submerged metal boiler cup (see figure 20.1), but not so much water that it could slosh into the cup when the water is boiling. Start heating the water to a full boil as you proceed to the next steps.

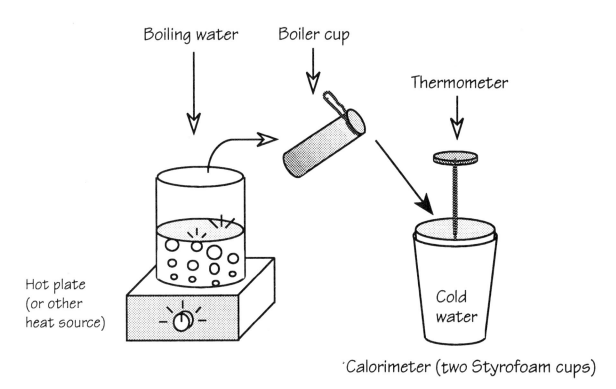

Figure 20.1

3. Measure and record the mass of a dry boiler cup. Pour metal shot into the boiler cup until it is about one-third filled, then measure and record the mass of the cup plus shot. Record the mass of the metal sample (m_s) in Data Table 20.1.

4. Carefully insert a thermometer into the metal shot, positioning it so the sensing end is in the middle of the shot, not touching the sides of the boiler cup. Carefully lower the boiler cup into the boiling water. Heat the metal shot until it is in the range of 90° to 95° C. Allow the sample to continue heating as you prepare the water and calorimeter cup (steps 5 and 6).

5. Acquire or make a calorimeter cup of two Styrofoam cups, one placed inside the other (figure 20.1) to increase the insulating ability of the cup. Measure and record the mass of the two cups. Add just enough water to the cup to cover the metal shot when it is added to the cup. This water should be cooler than room temperature (this is to balance possible heat loss by radiation). Measure and record the initial temperature of the water (T_{iw}) in Data Table 20.1.

6. Determine the mass of the cup with the water in it, then subtract the mass of the cup to find the mass of the cold water (m_w). Record the mass of the cold water in Data Table 20.1.

7. Measure and record the temperature of the metal shot. Record the initial temperature of the sample (T_{is}) in Data Table 20.1.

8. Pour the metal shot into the the water in the Styrofoam calorimeter cup. Stir and measure the

temperature of the mixture until the temperature stabilizes. Record this stabilized temperature and the final temperature for the water (T_{fw}) and the final temperature for the metal sample (T_{fs}). Calculate the specific heat (c_s) of the metal sample. Note that ΔT_w is obtained from $|T_{fw} - T_{iw}|$ and ΔT_s is obtained from $|T_{fs} - T_{is}|$.

9. Repeat the above steps for sample 2, recording all measurement data in Data Table 20.2. Repeat the procedure for sample 3, recording all measurement data in Data Table 20.3. Run a second trial on all three samples, comparing the results of both trials on each sample. Compare the calculations from the two trials on each sample to decide if a third trial is needed.

Results

1. Calculate the specific heat (c_s) for each sample. Show all work and record your result in each data table.
2. Using the accepted value for each sample, calculate the percentage error and record it in each data table.
3. Discuss and evaluate the magnitude of various sources of error in this experiment.

4. What would happen to the calculated specific heat if some boiling water were to slosh into the cup with the metal?

5. Was the purpose of this lab accomplished? Why or why not? (Your answer to this question should show thoughtful analysis and careful, thorough thinking.)

Data Table 20.1   Specific Heat of _____

	Trial 1	Trial 2
Mass of Sample (m_s)		
Initial Temperature of Cold Water (T_{iw})		
Mass of Cold Water (m_w)		
Initial Temperature of Metal Sample (T_{is})		
Final Temperature of Metal Sample (T_{fs})		
Final Temperature of Water (T_{fw})		
Calculated Specific Heat (c_s)		
Accepted Value		
Percent Error		

Data Table 20.2   Specific Heat of _____

	Trial 1	Trial 2
Mass of Sample (m_s)		
Initial Temperature of Cold Water (T_{iw})		
Mass of Cold Water (m_w)		
Initial Temperature of Metal Sample (T_{is})		
Final Temperature of Metal Sample (T_{fs})		
Final Temperature of Water (T_{fw})		
Calculated Specific Heat (c_s)		
Accepted Value		
Percent Error		

Data Table 20.3   Specific Heat of _____

	Trial 1	Trial 2
Mass of Sample (m_s)		
Initial Temperature of Cold Water (T_{iw})		
Mass of Cold Water (m_w)		
Initial Temperature of Metal Sample (T_{is})		
Final Temperature of Metal Sample (T_{fs})		
Final Temperature of Water (T_{fw})		
Calculated Specific Heat (c_s)		
Accepted Value		
Percent Error		

Name_____Section_____Date_____

Experiment 21: Static Electricity

Introduction

Charges of static electricity are produced when two dissimilar materials are rubbed together. Often the charges are small or leak away rapidly, especially in humid air, but they can lead to annoying electrical shocks when the air is dry. The charge is produced because electrons are moved by friction and this can result in a material acquiring an excess of electrons and becoming a negatively charged body. The material losing electrons now has a deficiency of electrons and is a positively charged body. All electric static charges result from such gains or losses of electrons. Once charged by friction, objects soon return to the neutral state by the movement of electrons. This happens more quickly in humid air because water vapor assists with the movement of electrons from charged objects. In this experiment you will study the behavior of static electricity, hopefully on a day of low humidity.

Procedure

Part A: Attraction and Repulsion

1. Rub a glass rod briskly for several minutes with a piece of nylon or silk. Suspend the rod from a thread tied to a wooden meterstick as shown in figure 21.1. Rub a second glass rod briskly for several minutes with nylon or silk. Bring it near the suspended rod and record your observations in Data Table 21.1. (If nothing is observed to happen, repeat the procedure and rub both rods briskly for twice the time.)

2. Repeat the procedure with a hard rubber rod that has been briskly rubbed with wool or fur. Bring a second hard rubber rod that has also been rubbed with wool or fur near the suspended rubber rod. Record your observations as in procedure step 1.

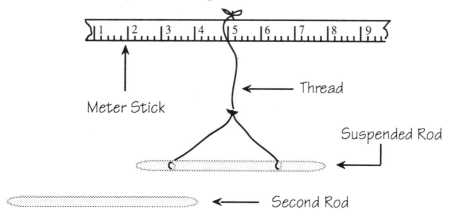

Figure 21.1

3. Again rub the hard rubber rod briskly with wool or fur and suspend it. This time briskly rub a glass rod with nylon or silk and bring the glass rod near the suspended rubber rod. Record your observations.

4. Briskly rub a glass rod with nylon or silk and bring it near, but not touching, the terminal of an electroscope (figure 21.2). Record your observations.

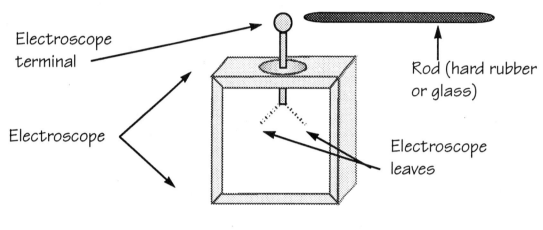

Figure 21.2

5. Repeat procedure step 4 with a hard rubber rod rubbed with wool or fur, again not touching the electroscope terminal. Record your observations.

Part B: Charging by Induction

1. Inflate two rubber balloons and tie the ends. Attach threads to each balloon and hang them next to each other from a support. Rub both balloons with fur or wool and allow them to hang freely. Record your observations in Data Table 21.2.

2. Bring a hard rubber rod that has been rubbed with wool or fur near the rubbed balloons. Record your observations.

3. Bring a glass rod that has been rubbed with nylon or silk near the rubbed balloons. Record your observations.

4. Detach one of the balloons by breaking or cutting the thread. Rub the balloon with fur or wool for several minutes. Hold the balloon against a wall and slowly release it. Record your observations.

5. Move the rubbed balloon near an electroscope and record your observations.

6. Move an electroscope near the wall where the balloon was held. Record your observations.

Part C: Determining the Sign of a Charge

1. When a rubbed hard rubber rod is brought near the terminal of an electroscope the leaves will stand apart but fall back together when the rod is removed.

2. When a rubbed hard rubber rod touches the terminal of an electroscope the leaves stand apart as before. When the rod is removed this time the leaves *remain* apart.

3. When the charged rod was brought near the terminal a charge was *induced* by the reorientation of charges in the terminal and leaves. When the rod was removed, the charges returned to their original orientation and the leaves collapsed because no net charge remained on the electroscope.

4. When the electrode was touched, charge was transferred to (or from) the electroscope and removing the rod had no effect on removing the charge. Touching the terminal with your finger returns the electroscope to a neutral condition.

5. An electroscope may be used to determine the sign of a charged object. First, charge the electroscope by induction as in procedure step 3 above. While the charged rod is near the terminal, touch the opposite side of the terminal with a finger of your free hand. Electrons will be repelled and conducted away through your finger. Remove your finger from the terminal, then move the rubber rod from near the electroscope. The electroscope leaves now have a net positive charge. If a charged object is brought near the electroscope the leaves will spread farther apart if the object has a positive charge. If the charged object has a negative charge, electrons are repelled into the leaves and they will move together as they are neutralized.

6. The process of an object gaining an excess of electrons or losing electrons through friction is complicated and not fully understood theoretically. It is possible experimentally, however, to make a list of materials according to their ability to lose or gain electrons. Gather various materials such as polyethylene film, rubber, wood, cotton, silk, nylon, fur, wool, glass, and plastic. Give an electroscope a positive charge by induction as described in procedure step 5. Rub combinations of the materials together and determine if the charge on each material is positive or negative. Record your findings.

Results

1. Describe two different ways that electrical charge can be produced by friction.

2. Describe how you can determine the sign of a charged object. What assumption must be made using this procedure?

3. Move a hard rubber rod that has been rubbed with wool or fur near a very thin, steady stream of water from a faucet. Describe, then explain your observations.

4. Was the purpose of this lab accomplished? Why or why not? (Your answer to this question should be reasonable and make sense, showing thoughtful analysis and careful, thorough thinking.)

Going Further

Make a charge detector. Spray two grains of puffed rice or wheat with a very thin layer of aluminum paint. Use a needle and thread to suspend the grains from a stopper or cork. Use this charge detector to investigate charged plastic rods, combs, glass rods, and other items.

Data Table 21.1   Attraction and Repulsion of Glass Rod and Rubber Rod

Interaction	Observations
Glass rod - Glass rod	
Rubber rod - Rubber rod	
Glass rod - Rubber rod	
Glass rod - Electroscope	
Rubber rod - Electroscope	

How many kinds of electric charge exist according to your findings above? Explain your reasoning.

How do charges interact?

Data Table 21.2   Charging by Induction

Interaction	Observations
Balloon - Balloon	
Rubber rod - Balloon	
Glass rod - Balloon	
Balloon - Wall	
Balloon - Electroscope	
Wall - Electroscope	

What evidence did you find to indicate that the balloons had static charges?

Describe the evidence you found to indicate that the wall was or was not charged as shown by the electroscope. Explain.

Explain why a balloon exhibits the behavior that it did on the wall.

Name_____Section_____Date_____

Experiment 22: Electric Circuits

Introduction

An electric *circuit* is a conducting path for an electric *current*, which is a flow of charge. When the circuit is connected to a battery, for example, charges flows from one terminal of the battery, through one or more electrical devices, and then back to the other terminal. Circuits can be described with words or by symbols that are widely known and used. Some of these symbols are illustrated in figure 22.1.

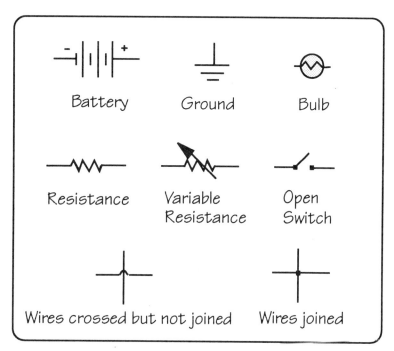

Figure 22.1

Procedure

1. Use *one* flashlight battery, *one* flashlight bulb, and *one* 10 cm piece of hook-up wire to make the bulb light. On page 198, make a sketch of your circuit. Use the symbols given in the introduction to show exactly how you constructed the circuit to make the bulb light. Draw an arrow to show the flow of **conventional current** and a second arrow to show the flow of the **electron current**. Identify both arrows.
2. Note the construction of the flashlight bulb. Figure 22.2 is an incomplete diagram of a bulb. Complete this diagram by drawing wires that connect the filament to the contact points in such a way that there is a complete circuit for the current to follow.
3. Wire the circuits that are shown in figure 22.3. For each circuit, note if a single bulb lights dimly, brightly, or if at all, and when two bulbs are lit note if unscrewing one affects the other in any way. Record these and other observations in Data Table 22.1.

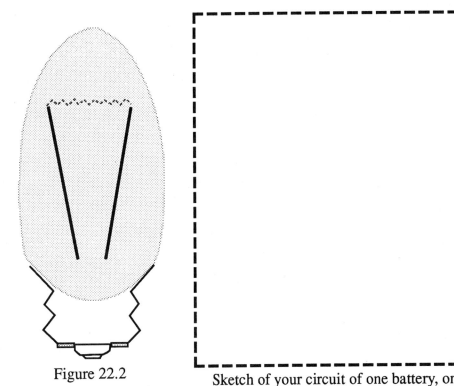

Figure 22.2

Sketch of your circuit of one battery, one bulb, and one wire.

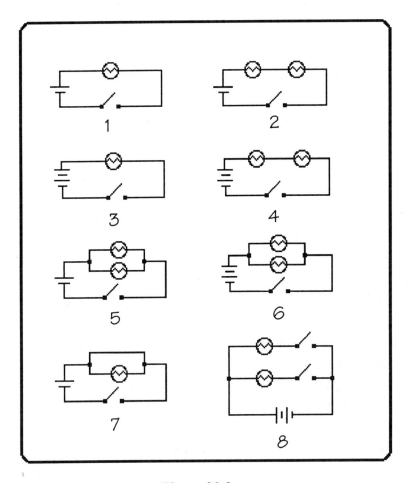

Figure 22.3

Results

1. Using the brightness of the bulb in the simple circuit (#1) as the standard for normal, describe which circuits had bulbs with a normal brightness. (In your attempt to control variables, note the color of the small glass blob at the base of the filament in each bulb. Different colors indicate bulbs of different resistance.)

2. Which circuits had bulbs that were dimmer than normal? Explain.

3. Which circuits had bulbs that were brighter than normal? Explain.

4. In which circuits did removing one bulb cause the other to go out? Explain.

5. In which circuits did removing one bulb not affect the other? Explain.

6. Describe the general requirements needed for any circuit to function.

7. Was the purpose of this lab accomplished? Why or why not? (Your answer to this question should show thoughtful analysis and careful, thorough thinking.)

Data Table 22.1	Circuit and Bulb Brightness Relationships
Circuit	Observations
Circuit 1, a simple circuit.	
Circuit 2, two bulbs in series.	
Circuit 3, two batteries in series.	
Circuit 4, two bulbs in series with two batteries in series.	
Circuit 5, two bulbs in parallel.	
Circuit 6, two bulbs in parallel and two batteries in series.	
Circuit 7, one bulb in parallel with wire. Show with arrows how current takes path of least resistance.	
Circuit 8, switches in series with bulbs in parallel circuit.	

Name_____Section_____Date_____

Experiment 23: Ohm's Law

Introduction

An electric charge has an electric field surrounding it, and work must be done to move a like-charged particle into this field since like charges repel. The electrical potential energy is changed just as gravitational potential energy is changed by moving a mass in the earth's gravitational field. A charged particle moved into the field of a like-charged particle has potential energy in the same way that a compressed spring has potential energy. In electrical matters the potential difference that is created by doing work to move a certain charge creates electrical potential. A measure of the electrical potential difference between two points is the **volt** (V).

A volt measure describes the potential difference between two places in an electric circuit. By analogy to pressure on water in a circuit of water pipes the potential difference is sometimes called an "electrical force" (emf). Also by analogy to water in a circuit of water pipes, there is a varying rate of flow at various pressures. An electric **current** (I) is the quantity of charge moving through a conductor in a unit of time. The unit defined for measuring this rate is the **ampere** (A), or the **amp** for short.

The rate of water flow in a pipe is directly proportional to the water pressure; e.g., a greater pressure produces a greater flow. In an electric circuit the current is directly proportional to the potential difference (V) between two points. Most materials, however, have a property of opposing or reducing a current, and this property is called **electrical resistance** (R). If a conductor offers a small resistance less voltage would be required to push an amp of current through the circuit. On the other hand, a greater resistance requires more voltage to push the same amp of current through the circuit. Resistance (R) is therefore a *ratio* of the potential difference (V) between two points and the resulting current. This ratio is the unit of resistance and is called an **ohm** (Ω). Another way to show the relationship between the voltage, current, and resistance is

$$R = \frac{V}{I}$$

or

$$V = IR$$

which is known as **Ohm's law**. This is one of the three ways to show the relationship; this one (solved for V) happens to be the equation of a straight line with a slope R when V is on the *y*-axis, I is on the *x*-axis, and the *y*-intercept is zero.

Procedure

Part A: Known Resistance

1. A known resistance will be provided for use in this circuit.

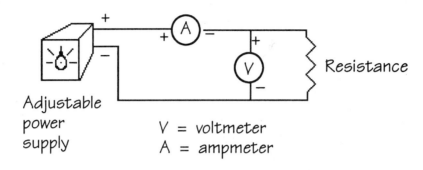

Figure 23.1

2. You will adjust the dc adjustable power supply as instructed by your laboratory instructor, obtaining six values for voltage and current using the supplied resistor. Set up the circuit with the power *off* and do not proceed until the laboratory instructor has checked the circuit.

3. Record the value of the resistor and the six values for the current and voltage in Data Table 23.1.

Part B: Unknown Resistance

Repeat procedure A with an unknown resistor. Record your data in Data Table 23.2.

Results

1. Make a graph of the six data points of Data Table 23.1, placing the current on the *x*-axis and the voltage on the *y*-axis. Calculate the slope and write it here and somewhere on the graph.

2. Compare the calculated value of the known resistor with the accepted value as given by your instructor. Calculate the percentage error.

3. Make a second graph, this time of the six data points in Data Table 23.2, again placing the current on the x-axis and the voltage on the y-axis. Calculate the slope and write it here and somewhere on the graph.

4. What is the value of the unknown resistor?

5. Explain how the two graphs demonstrate Ohm's law.

6. Was the purpose of this lab accomplished? Why or why not? (Your answer to this question should show thoughtful analysis and careful, thorough thinking.)

Going Further

1. Check your answer about the value of the unknown resistor by using your calculated value in the equation of a straight line when V = 2 V, 4 V, and 6 V. Verify with the laboratory equipment and calculate the average percentage error. Describe your results here:

2. Use three different resistances (e.g., 16 Ω, 30 Ω, and 47 Ω) connected in a series for four different input voltages (2 V, 4 V, 6 V, and 8 V) and connected in a parallel circuit. Plot voltage versus total current for both the series and parallel circuits and quantitatively show how the total resistance (the slope) differs for series and parallel circuits.

Data Table 23.1   Voltage and Current Relationships With Known Resistance

Trial	Voltage (V)	Current (I)
1	_____	_____
2	_____	_____
3	_____	_____
4	_____	_____
5	_____	_____
6	_____	_____

Resistor _____ Ω

Data Table 23.2   Voltage and Current Relationships With Unknown Resistance

Trial	Voltage (V)	Current (I)
1		
2		
3		
4		
5		
6		

Resistor _____ Ω

Name_____Section_____Date_____

Experiment 24: Series and Parallel Circuits

Introduction

There are two basic ways to connect more than one resistance in a circuit, in a *series circuit* or in a *parallel circuit*. A **series circuit** has each resistance connected one after the other so the same charges flow through one resistance, then the next one, and so on. A **parallel circuit** has separate pathways for the charges so they do not go through one resistance after the other. The use of the term "parallel" means that charges can flow through one of the branches, but not the others. The term does not mean that the branches are necessarily lined up with each other. The series and parallel circuits have separate characteristics that offer certain advantages and disadvantages.

Adding more resistances in a series circuit results in two major effects that are characteristic of all series circuits. More resistances result in (1) a *decrease in the current* available in the circuit, and (2) a *reduction of the voltage* available for each resistance. Since power is determined from I × V, adding more lamps will result in dimmer lights. Perhaps you have observed such a dimming when you connected two strings of Christmas tree lights. Many Christmas tree lights are connected in a series circuit. Another disadvantage to a series circuit is that if one bulb burns out the circuit is broken and all the lights go out.

Adding more resistances in a parallel circuit results in three major effects that are characteristic of all parallel circuits. More resistances result in (1) an *increase in the current* in the circuit, (2) as *the same voltage* is maintained across each resistance, and (3) *lowering of the total resistance* of the entire circuit. The total resistance is lowered since additional branches provide more pathways for the charges to move.

In this investigation you will investigate some of the characteristics of series and parallel circuits as you measure the potential difference, current, and resistance and calculate the power of the same bulb in both circuits.

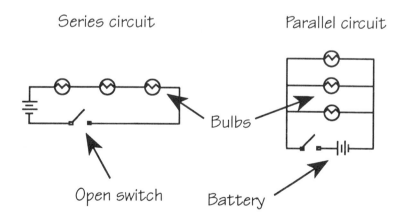

Figure 24.1

Procedure

You will be using three #41 bulbs (0.5 A), bulb sockets, two dry cells (or other source of 3 V dc), a dc voltmeter (0 to 5 V range), and a dc ammeter (0 to 2 A range) to read potential difference and current values at various points in a series circuit and in a parallel circuit. Note that an ammeter is always connected in series in a circuit so all the desired current passes through it. A voltmeter is always wired in parallel so that it measures the difference in potential between two points in the circuit. Also note that one of the terminals of the dc meter is marked negative (–). Always connect this negative (–) terminal of the meter to the negative (–) terminal of a battery (or other voltage source). If you connect the (–) of the meter to the (+) of a voltage source the pointer of the meter will move backward off the dial, which could damage the meter.

1. Connect two dry cells in series, then wire the connected cells to three lamps and a switch in series as shown in figure 24.2.

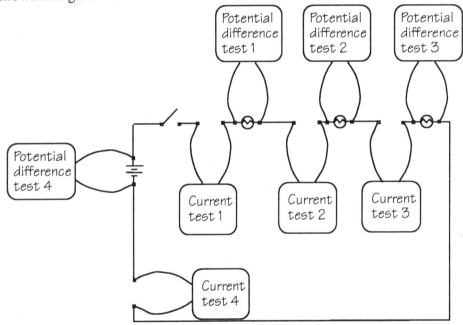

Figure 24.2

2. Make four separate voltage readings with a voltmeter at the places in the circuit as shown in figure 24.2. Note that test 1 measures the voltage drop for bulb 1, test 2 measures the voltage drop for bulb 2, test 3 measures the voltage drop for bulb 3, and test 4 measures the voltage source for the entire circuit. Record your findings in Data Table 24.1.

3. Make four separate current readings with an ammeter at the places in the circuit as shown in figure 24.2. Note that test 1 measures the current before bulb 1, test 2 measures the current before bulb 2, test 3 measures the current before bulb 3, and test 4 measures the current for the entire circuit. Record your findings in Data Table 24.1.

4. Remove one of the bulbs from the circuit and record here the effect on the other lamps:

5. Connect the three lamps in parallel as shown in figure 24.3.

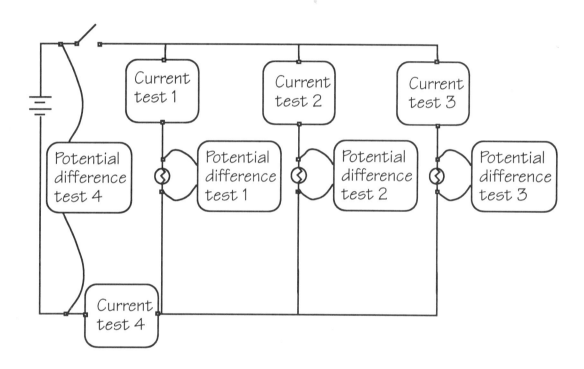

Figure 24.3

6. Make four separate voltage readings with a voltmeter at the places in the circuit as shown in figure 24.3. Note that test 1 measures the voltage drop for bulb 1, test 2 measures the voltage drop for bulb 2, test 3 measures the voltage drop for bulb 3, and test 4 measures the voltage source for the entire circuit. Record your findings in Data Table 24.2.

7. Make four separate current readings with an ammeter at the places in the circuit as shown in figure 24.3. Note that test 1 measures the current before bulb 1, test 2 measures the current before bulb 2, test 3 measures the current before bulb 3, and test 4 measures the current for the entire circuit. Record your findings in Data Table 24.2.

8. Remove one of the lamps from the circuit and record here the effect on the other lamps:

9. In Data Table 24.1, calculate and record the resistance and power of each bulb and for the entire series circuit.

10. In Data Table 24.2, calculate and record the resistance and power of each bulb and for the entire parallel circuit.

Results

1. How does the voltage drop of each of the three bulbs in the series circuit compare with the voltage of the source?

2. How does the voltage drop of each of the three bulbs in the parallel circuit compare with the voltage of the source?

216

3. How does the current through one bulb in the series circuit compare with the total current through the three lamps?

4. How does the current through one bulb in the parallel circuit compare with the total current through the three lamps?

5. Account for any differences in brightness observed by the same bulbs in series and parallel circuits.

6. According to the results of this investigation, what happens to the current and voltage available for each resistance as more are added (a) to a series circuit, (b) to a parallel circuit?

7. Other than differences in current, voltage, and resistance, what distinguishing characteristic will tell you if a circuit is a series or parallel circuit?

8. Was the purpose of this lab accomplished? Why or why not? (Your answer to this question should show thoughtful analysis and careful, thorough thinking.)

Going Further

What are the advantages and disadvantages of using either a series or parallel circuit for household wiring? Include possibilities for increased load, resistance to electrical work, and safety considerations in your analysis.

Data Table 24.1   Series Circuit

Bulb or Circuit	Volts (V)	Amps (A)	Resistance (Ω)	Power (W)
1				
2				
3				
Whole Circuit				

Data Table 24.2	Parallel Circuit			
Bulb or Circuit	Volts (V)	Amps (A)	Resistance (Ω)	Power (W)
1				
2				
3				
Whole Circuit				

Name_____Section_____Date_____

Experiment 25: Magnetic Fields

Introduction

A magnet moved into the space near a second magnet experiences a force as it enters the **magnetic field** of the second magnet. The magnetic field model is a conceptual way of considering how two magnets interact with one another. The magnetic field model does not consider the force that one magnet exerts on another one through a distance. Instead, it considers *the condition of space around a magnet*. The condition of space around a magnet is considered to be changed by the presence of the magnet. Since this region of space, or field, is produced by a magnet, it is called a *magnetic field*. A magnetic field can be represented by *magnetic field lines*. By convention, magnetic field lines are drawn to indicate how the *north pole* of a tiny imaginary magnet would point when in various places in the magnetic field. Arrowheads indicate the direction that the north pole would point, thus defining the direction of the magnetic field. The strength of the magnetic field is greater where the lines are closer together and weaker where they are farther apart. Magnetic field lines emerge from a magnet at the north pole and enter the magnet at the south pole. Magnetic field lines always form closed loops.

Magnetic field strength is defined in terms of the magnetic force exerted on a test charge of a particular charge and velocity. The magnetic field is thus represented by vectors (symbol B) that define the field strength, also called the magnetic induction. The units are:

$$B = \frac{\text{newton}}{(\text{coulomb})\left(\frac{\text{meters}}{\text{second}}\right)}$$

Since a coulomb/s is an amp, this can be written as

$$B = \frac{\text{newton}}{\text{amp} \cdot \text{meter}}$$

which is called a **tesla** (T). The tesla is a measure of the strength of a magnetic field. Near the surface, the earth's horizontal magnetic field in some locations is about 2×10^{-5} tesla. A small bar magnet produces a magnetic field of about 10^{-2} tesla but large, strong magnets can produce magnetic fields of 2 tesla. Superconducting magnets have magnetic fields as high as 30 tesla. Another measure of magnetic field strength is called the **gauss** (G) (1 tesla = 10^4 gauss). Thus the process of demagnetizing something is sometimes referred to as "degaussing."

In this experiment you will investigate the magnetic field around a permanent magnet and the magnetic field around a current-carrying conductor.

Procedure

1. Tape a large sheet of paper on a table, with the long edge parallel to the north-south magnetic direction as determined by a compass.

2. Center a bar magnet on the paper with its south pole pointing north. Use a sharp pencil to outline lightly the bar magnet on the paper. Write N and S on the paper to record the north-seeking and south-seeking poles of the bar magnet. Place the bar magnet back on its outline if you moved it to write the N and the S.

3. Slide a small magnetic compass across the paper, stopping close to the north-seeking pole of the bar magnet. Make two dots on the paper, one on either side of the compass and aligned with the compass needle. See figure 25.1.

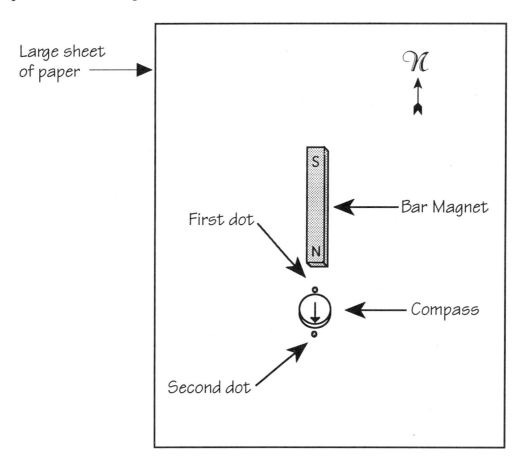

Figure 25.1

4. Slide the compass so the south pole of the needle is now directly over the dot that was at the north pole of the needle. Make a new dot at the north pole end of the compass, exactly in front of the needle. See figure 25.2.

222

5. Continue the process of moving the compass so the south pole of the needle is over the most recently drawn dot, then make another dot at the north pole of the needle. Stop when you reach the bar magnet or the edge of the paper.

6. Draw a smooth curve through the dots, using several arrowheads to show the direction of the magnetic flux line.

7. Repeat procedure steps 3 through 6, by starting with the compass in a new location somewhere around the bar magnet. Repeat the procedures until enough flux lines are drawn to make a map of the magnetic field.

8. Place a second large sheet of paper on a large rigid plastic sheet (or glass plate) on top of the bar magnet. Sprinkle a thin, even layer of iron filings over the plastic, tapping the sheet lightly as you sprinkle. Sketch the magnet flux lines on the paper as shown by the arrangement of the filings.

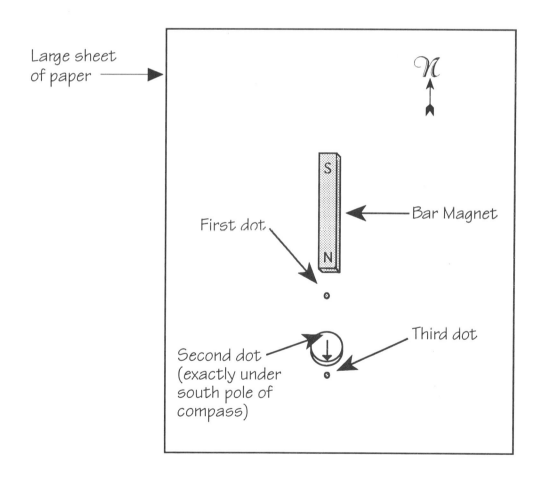

Figure 25.2

Results

1. How is using iron filings (a) similar, and (b) different, from using a magnetic compass to map a magnetic field?

2. In terms of a force, or torque on a magnetic compass needle, what do the lines actually represent? Explain.

3. Do the lines ever cross each other at any point? Explain.

4. Where do the lines appear to be concentrated the most? What does this mean?

Name_____ Section_____ Date_____

Experiment 26: Electromagnets

Introduction

Electric charges in motion produce a magnetic field around the charges, and a current-carrying wire has a magnetic field as a result. You can "map" such a magnetic field by running a straight wire vertically through a sheet of paper. The wire is connected to a battery and iron filings are sprinkled on the paper. The filings will become aligned as the domains in each tiny piece of iron are forced parallel to the field. Overall, filings near the wire form a pattern of concentric circles with the wire in the center.

The direction of the magnetic field around a current-carrying wire can be determined by using the common device for finding the direction of a magnetic field, the magnetic compass. The north-seeking pole of the compass needle will point in the direction of the magnetic field lines (by definition). If you move the compass around the wire the needle will always move to a position that is tangent to a circle around the wire. Evidently the magnetic field lines are closed, concentric circles that are at right angles to the length of the wire. If you *reverse* the direction of the current in the wire and again move a compass around the wire, the needle will again move to a position that is tangent to a circle around the wire. This time the north pole direction is reversed. Thus the magnetic field around a current-carrying wire has closed concentric field lines that are perpendicular to the length of the wire. The *direction* of the magnetic field is determined by the direction of the current.

The *strength* of the magnetic field (B) around a long, straight current-carrying wire is directly proportional to the current (I) in the wire and inversely proportional to the distance (d) from the wire, or

$$B = k\frac{2I}{d}$$

where the proportionality constant k is 1.00×10^{-7} newton/amp^2. Note that the magnetic field strength varies with the distance from the wire (not the square of the distance) and that the unit is newton/amp·meter, or tesla.

A current-carrying wire that is formed into a loop has perpendicular, circular field lines that pass through the inside of the loop in the same direction. This has the effect of concentrating the field lines, which increase the magnetic field intensity. Since the field lines all pass through the loop in the same direction, one side of the loop will have a north pole and the other side a south pole. Many loops of wire formed into a cylindrical coil is called a **solenoid**. When a current is in a solenoid, each loop contributes field lines along the length of the cylinder. The overall effect is a magnetic field around the solenoid that acts just like the magnetic field of a bar magnet. This magnet, called an **electromagnet**, can be turned on or off by turning the current on and off. In addition, the strength of the electromagnet depends on the magnitude of the current and the number of loops (ampere-turns). The strength of the electromagnet can also be increased by placing a piece of soft iron in the coil.

The magnetic domains of the iron become aligned by the influence of the magnetic field. This induced magnetism increases the overall magnetic field strength of the solenoid as the magnetic field lines are gathered into a smaller volume within the core.

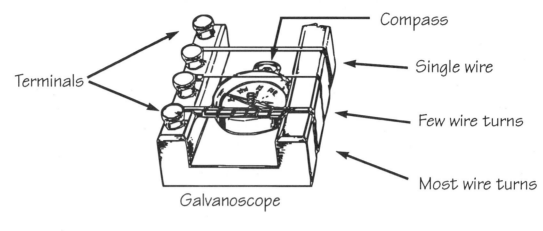

Figure 26.1

Procedure

1. Place a galvanoscope on a nonmetallic table with the wire coils running in a north-south direction. Place a compass beneath the coil of wires with the most turns. Situate the compass and the device so the needle is parallel and beneath the wire coil, with the north pole of the compass needle pointing to the zero degree mark.

2. Connect a dry cell or laboratory source of current at 1.5 V to the terminal of most turns. Note the direction of the current (north-to-south or south-to-north), the direction of the needle deflection (east or west), and the amount of deflection in degrees. Disconnect the source of current. Record your observations in Data Table 26.1.

3. Reverse the direction of the current and repeat the observations and recordings of procedure step 2. Disconnect the source of current, then record your observations in Data Table 26.1.

4. With the current disconnected, carefully move the compass to beneath the coil of wires with a few turns, otherwise situating the compass and device exactly as they were in procedure step 1. Repeat procedure steps 2 and 3 with the compass beneath this coil and record your observations in Data Table 26.1.

5. Repeat procedure step 4, this time moving the compass to beneath the single wire. Record all observations as before.

6. Obtain about 3 m of No. 18 insulated copper wire and a 1/2 cm diameter soft iron spike that is about 12 cm long. This is sufficient wire to leave about 20 cm free, then wrap about 100 turns around the spike, leaving about 20 cm free at the opposite end. Use this device and a dry cell or

laboratory source of current at 1.5 V to begin a series of experiments to find out what factors affect the strength and polarity of an electromagnet. The strength of the electromagnet could be measured in terms of how many paper clips or nails it will pick up. Be sure to record all procedures tried as well as the results, remembering that "no result" is a finding as well as more dramatic events.

Results

1. What determines the *direction* of a magnetic field around a current-carrying wire? Provide evidence for your answer.

2. What determines the *strength* of a magnetic field around a current-carrying wire? Provide evidence for your answer.

3. Which is stronger, an electromagnet with an iron core or an electromagnet without an iron core? Explain.

4. Was the purpose of this lab accomplished? Why or why not? (Your answer to this question should show thoughtful analysis and careful, thorough thinking.)

Going Further

Will one electromagnet attract or repel another electromagnet when there is a current in both coils? Test this idea with an experiment after recording your prediction.

Data Table 26.1	Direction and Strength of Magnetic Field Around Electromagnet		
Wire Coil	Direction of Current	Direction of Deflection	Amount of Deflection
Most Turns			
Most Turns (reverse current)			
Several Turns			
Several Turns (reverse current)			
Single Wire			
Single Wire (reverse current)			

Experiment 27: Speed of Sound in Air

Introduction

A vibrating tuning fork sends a series of condensations and rarefactions through the air. When the tuning fork is held over a glass tube that is closed at the bottom, the condensations and rarefactions are reflected from the bottom. At certain lengths of tube, the reflected condensations and rarefactions are in phase with those being sent out by the tuning fork and an increase of amplitude occurs from the resonant condition. Figure 27.1 shows a wave trace representing one wavelength in which the reflected wave is in phase with the incoming wave, forming a standing wave. The antinodes represent places of maximum vibration and increased amplitude.

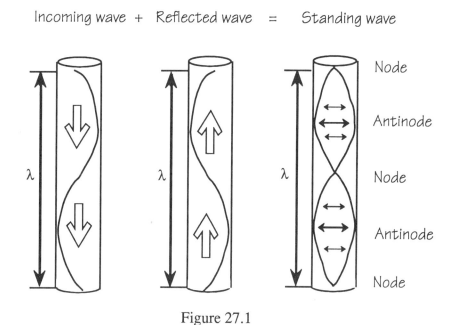

Figure 27.1

Resonance occurs when the length of the tube is such that an antinode (the place of maximum vibration) occurs at the open end. As you can see from the sketch above, there are two situations when this would occur for tube lengths less than one wavelength, 1/4 of the way up and 3/4 of the way up from the bottom. Thus resonance occurs when the length of the tube (L) is equal to 1/4 λ, 3/4λ, 5/4λ, and so forth where λ is the wavelength of the sound wave produced by the tuning fork.

In this experiment, a vibrating tuning fork is held just above a cylinder that is open at one end. The length to the closed end is adjusted by adding or removing water. The lowest frequency (the fundamental frequency) occurs when the longest wavelength has an antinode at the open end, so the length of the open tube is about 1/4 of the wavelength of the fundamental frequency as shown in figure 27.2. Since the length of the tube at this fundamental frequency is L = 1/4 λ, then the fundamental wavelength must be $\lambda = 4L$.

Using the wave equation
$$v_T = f\lambda$$
and substituting the known frequency of the tuning fork for f and the experimentally determined value for the wavelength λ, you can calculate the speed of sound v_T in the tube at room temperature by using the relationship

$$v_T = v_{0°C} + \left(\frac{0.6 \text{ m/s}}{°C}\right)(T_{room})$$

where $v_{0°C}$ is the speed of sound at 0° C (331.4 m/s) and T_{room} is the present room temperature in °C.

Procedure

1. The water level in the glass tube is adjusted by raising and lowering the supply tank. Adjust the tank so the glass tube is nearly full of water.

2. Strike the tuning fork with a rubber hammer and hold the vibrating tines just above the opening of the tube.

3. Lower the water level slowly while listening for the increase in the intensity of the sound that comes with resonance. Experiment with the *entire length of the tube*, seeing how many different places of resonance you can identify.

4. Using the information learned in procedure step 3, go to the resonance level immediately *below the resonance position* of the highest water level as shown in figure 27.2. (Make sure there is *not* another resonance point between the highest water level and this second level.) Slightly raise and lower the water level until you are sure that you have found the maximum intensity. Note the relationship between the wavelength and the length of the tube as shown in figure 27.2. Measure and record in Data Table 27.1 on page 234 the length of this resonating air column to the nearest millimeter. Change the water level and run two more trials, again locating the distance with the maximum sound. Record these two lengths in Data Table 27.1 and average the length for the three trials. Record the frequency of the tuning fork (usually stamped on the handle) and the room temperature.

5. Repeat procedure steps 1 through 4 for the second resonance point *at the highest water level,* with an air column about one-third the length of the first as shown in figure 27.3. (Again, make sure there is *not* another resonance point between the highest water level and this second level.) Note the relationship between the wavelength and the length of the tube as shown in figure 27.3. Run three trials at this position and record the data in Data Table 27.2 on page 234 and, as before, average the three trials. Record the frequency of the tuning fork and the room temperature (do not assume that the room temperature remains constant).

6. Repeat the entire procedure using a different tuning fork with a different frequency. Record all data in Data Tables 27.3 and 27.4 on page 235.

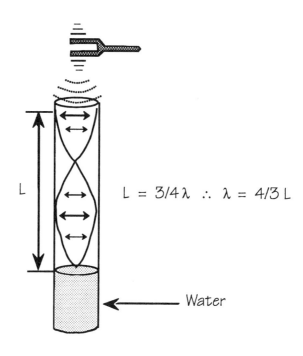

Figure 27.2

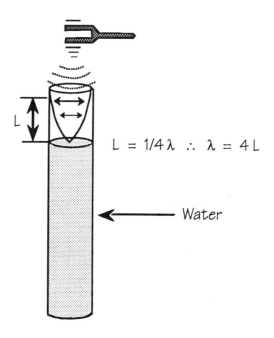

Figure 27.3

Results

1. Calculate the velocity of sound at room temperature for both tuning forks at both resonance positions and record in data tables at the measured room temperatures. Here, write the average values for both tuning forks:

2. Using the accepted value of sound in dry air at the measured room temperature, calculate the percentage error for both tuning forks [accepted value = 331.4 m/s + (0.6 m/s/°C)(T_{room})].

3. Analyze and discuss the possible sources of error in this experiment.

4. Describe how you could do a similar experiment to find the frequency of a tuning fork with an unknown frequency.

5. Was the purpose of this lab accomplished? Why or why not? (Your answer to this question should show thoughtful analysis and careful, thorough thinking.)

Data Table 27.1   Resonance in an Air Column: Lowest Position - Frequency 1

	Trial 1	Trial 2	Trial 3	Average
Length of Resonating Air Column (m)				
Room Temperature (°C)				

Calculated Wavelength (m) ....

Calculated Velocity in Air (m/s) ....

Tuning Fork Frequency (Hz) ....

Data Table 27.2   Resonance in an Air Column: Next Higher Position - Frequency 1

	Trial 1	Trial 2	Trial 3	Average
Length of Resonating Air Column (m)				
Room Temperature (°C)				

Calculated Wavelength (m) ....

Calculated Velocity in Air (m/s) ....

Tuning Fork Frequency (Hz) ....

Data Table 27.3	Resonance in an Air Column: Lowest Position - Frequency 2			
	Trial 1	Trial 2	Trial 3	Average
Length of Resonating Air Column (m)				
Room Temperature (°C)				
Calculated Wavelength (m)				
Calculated Velocity in Air (m/s)				
Tuning Fork Frequency (Hz)				

Data Table 27.4	Resonance in an Air Column: Next Higher Position - Frequency 2			
	Trial 1	Trial 2	Trial 3	Average
Length of Resonating Air Column (m)				
Room Temperature (°C)				
Calculated Wavelength (m)				
Calculated Velocity in Air (m/s)				
Tuning Fork Frequency (Hz)				

Name_____Section_____Date_____

Experiment 28: Standing Waves

Introduction

    A stringed musical instrument, such as a guitar, violin, or piano, has strings that are stretched between two fixed ends. When a string is plucked, bowed, hit, or otherwise disturbed, waves of many different frequencies will travel back and forth on the string, reflecting from the fixed ends. Many of these waves quickly fade away but certain frequencies **resonate**, setting up patterns of waves. Reflected waves interfere with incoming waves of the same frequency, making (1) stationary places of destructive interference called **nodes**, which show no disturbance, and (2) loops of constructive interference called **antinodes**. Antinodes form where the crests and troughs of the two wave patterns produce a disturbance that rapidly alternates upward and downward. The pattern of alternating nodes and antinodes does not move along the string and is thus called a **standing wave.** A standing wave *for one wavelength* will have three nodes, one at each end and one in the center, and will have two antinodes. Standing waves occur at the natural, or resonant, frequency of a string and are determined by the length, tension, and density (mass per unit length) of the string.

    Standing waves are produced when a condition of resonance exists between the natural frequency of a string and the frequency of a disturbance. At resonance, there is a particular wavelength (λ) that is directly proportional to the velocity of the wave along the string and is inversely proportional to the frequency, or $v = f\lambda$, where f is the frequency and v is the velocity. This velocity is determined by the tension in the string (F_T) and the density (D) or mass per unit length of the string. The velocity is related to the tension (F_T) and the density (D) by

$$v = \sqrt{\frac{F_T}{D}}.$$

Changing the tension thus changes the velocity. Changes in the velocity result in changes in the wavelength at a constant frequency since $v = f\lambda$. Therefore, changing the tension at a constant frequency will result in different numbers of standing waves as the conditions for resonance are met by a varying tension.

    In this experiment standing waves are set up in a stretched nylon string held under tension by masses ($F_T = mg$) at one end and a vibrator at the other end. The vibrator is connected to a 60-cycle alternating current, which drives the vibrator at the power-line frequency (60 Hz) or else twice that frequency (120 Hz) depending on the orientation of the vibrator relative to the string. Most of the time the frequency will be double the supply current, or 120 Hz. A stroboscope can be used to check the frequency if there is a question.

    The tension in the string (F_T) is measured from the masses ($F_T = mg$) suspended over a pulley with a weight hanger and is changed by adding or removing masses.

    The length of a wave (λ) is twice the distance (L) between two consecutive nodes in a standing wave, so $\lambda = 2L$.

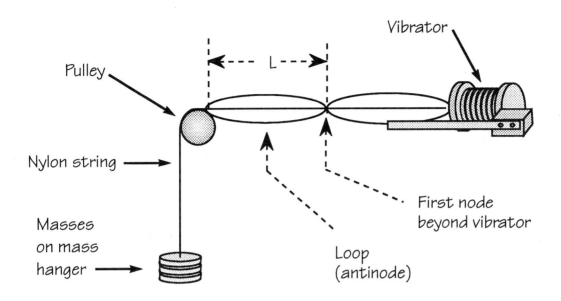

Figure 28.1

The relationship between the velocity (v), wavelength (λ), and frequency (f) is found in the wave equation, $v = f\lambda$. Solving for wavelength gives

$$\lambda = \frac{v}{f}.$$

But the velocity is related to the tension in the string (F_T) and the density (D) (i.e., mass per unit length) by

$$v = \sqrt{\frac{F_T}{D}}.$$

Substituting the tension and density variables for velocity in the wave equation gives a wavelength λ of

$$\lambda = \frac{\sqrt{\frac{F_T}{D}}}{f}$$

or

$$\lambda = \left(\frac{1}{f}\right)\left(\sqrt{\frac{F_T}{D}}\right).$$

At a constant frequency of vibration (f) and string density (D), the wavelength (λ) and the square root of the tension $\left(\sqrt{F_T}\right)$ should therefore be proportional, with a proportionality constant of

$$\frac{1}{\left(f\sqrt{D}\right)}.$$

Procedure

1. Obtain a length of nylon string of a known density, or measure a 150-cm length and determine the mass per unit length with a balance. In either case, record the density (D) or mass per unit length in the Data Table 28.1 in kg/m. Use one string only in this procedure.

2. Fasten one end of the string to the vibrator and pass the other end over a pulley about 1 meter away. Fasten a mass hanger to the end over the pulley (figure 28.1). Set the string length to 1.00 m.

3. Turn on the vibrator and add masses to the mass hanger until the string makes a standing wave of two definite segments. Adjust the tension by adding or removing small masses until the amplitude is at a maximum. If necessary, loosen the clamp holding the vibrator and carefully move it very slightly for maximum amplitude.

4. Measure the distance from the first node beyond the vibrator to the next node. The distance between two consecutive nodes is one-half wavelength, so two segments of measured length on the string equal one wavelength. Record the wavelength in meters and the tension in newtons for this resonant frequency.

5. Turn the vibrator on and adjust the tension by removing masses until standing waves of 3, 4, 5, 6, and 7 segments form for five separate trials. In each trial measure one segment and record the wavelength in meters with the corresponding tension in newtons.

Results

1. Plot the square root of the tension $\left(\sqrt{F_T}\right)$ in newtons $\left(\sqrt{N}\right)$ on the *x*-axis and the wavelength (λ) (in m) on the *y*-axis, using the entire graph paper. You should have a total of six data points.

2. Calculate the slope, showing your work here.

Should the slope have the same magnitude as $\dfrac{1}{\left(f\sqrt{D}\right)}$?

3. Determine the frequency (f) from the calculated slope and use this as the calculated average value for f.

4. Analyze the possible sources of error in this experiment.

5. What conclusions can you reach about the relationship between resonance and tension on a vibrating string?

6. What determines the quality of a musical note created by a vibrating string?

7. Was the purpose of this lab accomplished? Why or why not? (Your answer to this question should show thoughtful analysis and careful, thorough thinking.)

Going Further

1. Using unit analysis, show that $\dfrac{1}{(f\sqrt{D})}$ is the unit of the slope when λ is plotted against $\left(\sqrt{F_T}\right)$.

2. Describe an experiment that would show that the velocity of the wave (v) is proportional to $\sqrt{\left(\dfrac{F_T}{D}\right)}$. How is this v related to the frequency?

Data Table 28.1   Resonant Relationships of a String Under Tension

Number of Segments	Length of One Segment (m)	Wave Length (λ) (m)	Tension (F_T) (N)	Square Root of Tension ($\sqrt{F_T}$) ($\sqrt{N}$)
2	_____	_____	_____	_____
3	_____	_____	_____	_____
4	_____	_____	_____	_____
5	_____	_____	_____	_____
6	_____	_____	_____	_____
7	_____	_____	_____	_____

Density (D) (mass/length) _____ kg/m

Frequency (vibrations/sec) _____ Hz

Name_____Section_____Date_____

Experiment 29: Reflection and Refraction

Introduction

The travel of light is often represented by a **light ray**, a line that is drawn to represent the straight-line movement of light. The line represents an imaginary thin beam of light that can be used to illustrate the laws of reflection and refraction, the topics of this laboratory investigation.

A light ray travels in a straight line from a source until it encounters some object. What happens next depends on several factors, including the nature of the material making up the object, the smoothness of the surface, and the angle at which the light ray strikes the surface. If the surface is perfectly smooth, rays of light undergo **reflection.** if the material is transparent, on the other hand, the light ray may be transmitted through the material. In these cases the light ray appears to become bent, undergoing a change in the direction of travel at the boundary between two transparent materials (such as air and water). The change of direction of a light ray at the boundary is called **refraction**.

Light rays traveling from a source, before they are reflected or refracted, are called *incident rays*. If an incident ray undergoes reflection, it is called a *reflected ray*. Likewise, an incident ray that undergoes refraction is called a *refracted ray*. In either case, a line perpendicular to the surface, at the point where the incident ray strikes, is called the *normal*. The angle between an incident ray and the normal is called the *angle of incidence*. The angle between a reflected ray and the normal is called the *angle of reflection*. The angle between a refracted ray and the normal is called the *angle of refraction*. These terms are descriptive of their meaning, but in each case you will need to remember that the angle is measured from a line perpendicular to the surface, the **normal**.

Procedure

Part A: Reflection of Light

1. Using a ruler, draw a straight line across a sheet of plain (unlined) white paper. Place the paper on a smooth piece of cardboard that has been cut from a box. Label the line with a B at one end and B´ at the other end (B is for boundary).

2. Attach a small, flat mirror to a block of wood as shown in figure 29.1. Place the mirror and block combination on the paper with the back of the mirror (the reflecting surface) on line BB´.

3. Stick a pin straight up and down into the paper about 10 cm from the mirror and slightly to the

right side as shown in figure 29.1. On the left side, carefully align the edge of a ruler with the reflected image as shown in the illustration. Then firmly hold the ruler and draw a pencil line along this edge. Move the mirror and extend this line to the mirror boundary line BB´. Label the point of reflection with the letter P.

4. Place a protractor on line BB´ and mark a point 90° from the line. From this point, use the ruler to draw a dashed normal (NP). Complete your ray diagram by using the ruler to draw a line from the point of reflection (P) to the source of the light ray at the pin (I). Place arrows on line IP and line PR to show which way the light ray moved.

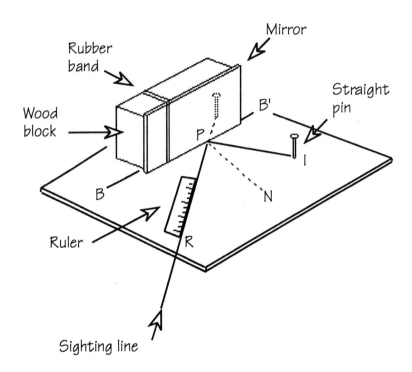

Figure 29.1

5. Use the protractor to measure the angle of incidence and the angle of reflection. Record these angles in Data Table 29.1 under Trial 1.

6. Place the mirror with its back edge again on line BB´ and conduct a second and third trial at different sighting angles. Record these measurements in Data Table 29.1 on page 250.

Part B: Refraction of Light

1. Place a clean sheet of white (unlined) paper on the cardboard. Place a glass plate approximately 5 cm square flat on the center of the paper. Use a pencil to outline the glass plate, then move the plate aside.

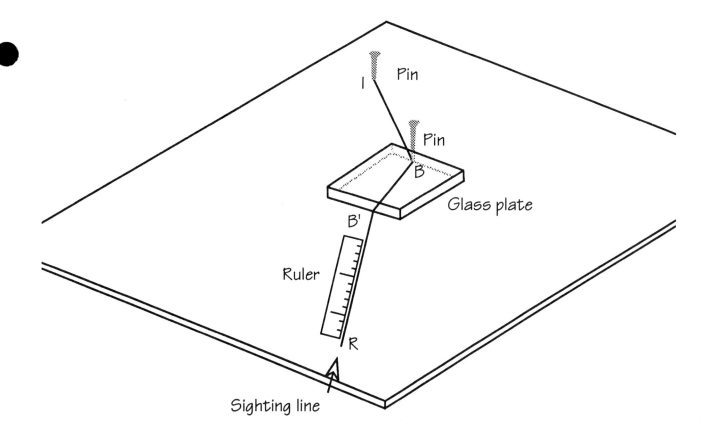

Figure 29.2

2. Use a ruler to draw a straight line from the upper left edge of the plate outline, making an angle of about 60° to the edge. Label this line IB as shown in figure 29.2. Place one upright pin at point B immediately outside the plate outline and a second upright pin at point I. Return the glass plate to the outline.

3. Bring the cardboard, paper, and glass plate near the edge of the table so you can sight through the glass plate toward the two pins. Position a ruler so that one edge aligns with the two pins as shown in figure 29.2. Draw a line along the ruler and label the line B´R. Move the glass plate aside for a second time.

4. Draw a line from B to B´, showing the path of the light ray through the glass. Overall, the path of the light ray is from IB to BB´ to B´R, showing that the light ray was bent twice.

5. Draw normals to the surface of the glass at B and B´. Show the angle of incidence and the angle of refraction with curved arrows at both boundaries.

Results

1. Describe any pattern you found in the data between the angle of incidence and the angle of reflection.

2. Describe what happens to a light ray as it travels (a) from air into glass and (b) from glass into air.

3. Assuming that light travels faster through air than it does through glass, make a generalized statement about what happens to a light ray with respect to the normal as it moves from a faster speed in one material to a slower speed in another.

4. What rules or generalizations do your findings suggest about reflection? How much more data would be required to make this a valid generalization?

5. What rules or generalizations do your findings suggest about refraction? How much more data would be required to make this a valid generalization?

6. Was the purpose of this lab accomplished? Why or why not? (Your answer to this question should be reasonable and make sense, showing thoughtful analysis and careful, thorough thinking.)

Going Further

Trace the travel of light rays through a convex lens, a concave lens, and a triangular prism. What are at least two factors that determine how much the light ray is bent in each lens?

Data Table 29.1	Reflection of Light	
Trial	Angle of Incidence	Angle of Reflection
1	_____	_____
2	_____	_____
3	_____	_____

Name_____Section_____Date_____

Experiment 30: Lenses

Introduction

A convex lens can be used as a "burning glass" by moving the lens back and forth until the sunlight is focused into a small bright spot—an image of the sun. This image is hot enough to scorch the paper, perhaps setting it on fire. The lens is moved back and forth to refract the parallel rays of light from the sun to the point where the image is formed on the paper. The place where the image forms is called the **focal point** of the lens. The distance from the focal point to the lens is called the **focal length** (f). The focal length of a lens is determined by its index of refraction and the shape of the lens. The focal length is an indication of the refracting ability, or strength, of a lens. A lens with a short focal length is considered to be a stronger lens than one with a longer focal length.

There are three important measurements that are used to describe how lenses work as optical devices. These are (1) the *focal length* (f), (2) the *image distance* (d_i), the distance from the lens that an image is formed, and (3) the *object distance* (d_o), the distance from the object being imaged to the lens. The relationship between these measurements is given in the **lens equation**, which is

$$\frac{1}{f} = \frac{1}{d_o} + \frac{1}{d_i}$$

The magnification produced by a lens is defined as the ratio of the height of the image (h_i) to the height of the object (h_o). This is also equal to the ratio of the image distance (d_i) to the object distance (d_o), or

$$\text{Magnification} = \frac{d_i}{d_o}$$

In this investigation you will compare measuring the focal length of a lens directly with using the lens equation to calculate the focal length of a convex lens. Magnification of a lens will also be investigated by comparing direct measurement of magnification with theoretical magnification as calculated from the focal lengths of two lenses.

Procedure

1. Measure the focal length of a lens directly. First, place a lens in a lens holder and secure it at the 50 cm mark on a meterstick. Second, point the meterstick at some distant objects, such as a tree or a house about a block away. Move a cardboard screen in a holder back and forth until you obtain a sharp image on the screen. Finally, measure the distance between the sharp image and the lens, which is the focal length (f) of the lens. Record the focal length in Data Table 30.1 on page 256.

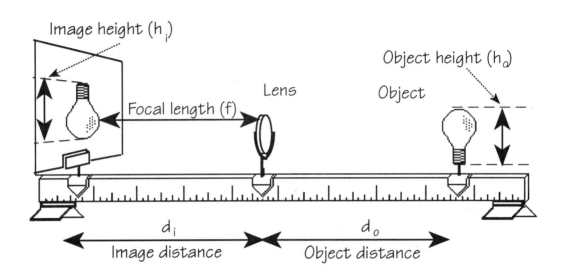

Figure 30.1

2. Find the focal length of the lens used in procedure step 1 by use of the lens equation. Set up a meterstick in holders, a screen in a holder, and a luminous object in a holder as shown in figure 30.1. The room should be darkened, then place the screen at the focal length distance found in procedure step 1. With the screen and object fixed in place, slowly move the lens along the meterstick to obtain the sharpest image possible. Note that the object and image should lie on a straight line along, and perpendicular to the principal axis of the lens. Measure and record in Data Table 30.1 the object distance (d_o) and the image distance (d_i) to the nearest 1 mm. Measure and record to the nearest 0.5 mm the height of the object (h_o) and height of the image (h_i). Record other observations here:

3. Place the lens *less than one focal length* from the bulb, then move the screen back and forth to see if you can obtain an image on the screen. Look through the lens at the bulb. Record your observations here:

4. Repeat procedure steps 1 and 2 for three more lenses. Record all data and results of calculations in Data Table 30.1. Select one of the lenses with a short focal length (for eyepiece lens) and one with a longer focal length (for objective lens) for use in the next procedure step.

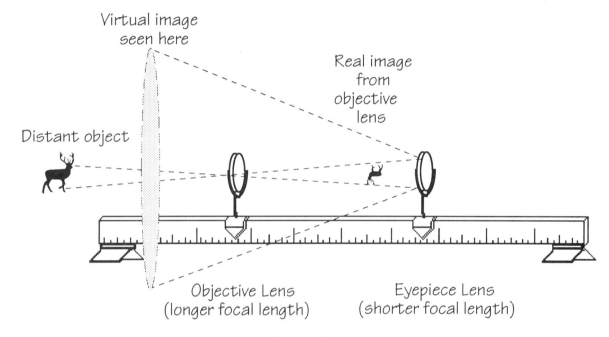

Figure 30.2

5. Make a telescope by mounting a short and longer focal length lenses on a meterstick as shown in figure 30.2. Focus the telescope by adjusting the shorter focal length lens until it magnifies the image from the longer focal length lens. Calculate the *theoretical* magnification of your telescope by dividing the focal length of the objective lens by the focal length of the eyepiece lens. Record the theoretical magnification in Data Table 30.2.

6. Work with a partner to *determine experimentally* the magnification of your telescope. Focus your telescope on some object in front of your partner, who should hold a meterstick next to the object. Look at the object normally with one eye and look at the object through the telescope with the other eye. Direct your partner to position a pointer on the meterstick to indicate the apparent size of the enlarged image. Record the height of the image and the height of the object in Data Table 30.2, then calculate the magnification.

Results

1. What relationships did you find between d_o, d_i, and f?

2. What relationships did you find between d_o, d_i, h_i, and h_o?

3. Discuss the advantages, disadvantages, and possible sources of error involved in the two ways of finding the focal length of lenses.

4. Discuss the advantages, disadvantages, and possible sources of error involved in the two ways of finding the magnification of lenses.

5. Was the purpose of this lab accomplished? Why or why not? (Your answer to this question should show thoughtful analysis and careful, thorough thinking.)

Going Further

Design an experiment to study the effect of the diameter of a lens on the image formed. Do the experiment.

Lens	Focal Length Measured Directly (f)	Object Distance (d_o)	Image Distance (d_i)	Focal Length from Lens Equation (f)	Object Size (h_o)	Image Size (h_i)	Magnification
1							
2							
3							
4							

Data Table 30.1 Lens Focal Length and Magnification

Data Table 30.2 Lens Magnification	
Objective Lens Focal Length	
Eyepiece Lens Focal Length	
Theoretical Magnification (focal length of objective ÷ focal length of eyepiece)	
Object Height (h_o)	
Image Height (h_i)	
Experimental Magnification ($h_i \div h_o$)	

Name_____Section_____Date_____

Experiment 31: Amount of Water Vapor in the Air

Introduction

The amount of water vapor in the air is referred to generally as **humidity**. A measurement of the amount of water vapor in the air at a particular time is called the **absolute humidity**. At room temperature, for example, humid air might contain 15 grams of water vapor in each cubic meter of air. At the same temperature air of low humidity might have an absolute humidity of only 2 grams per cubic meter. Absolute humidity can range from near zero up to a maximum that is determined by the temperature at a particular time, as shown in figure 31.1. Since the temperature of the water vapor present in the air is the same as the temperature of the air, the maximum absolute humidity is usually said to be determined by the air temperature. What this really means is that the maximum absolute humidity is determined by the temperature of the water vapor; that is, the average kinetic energy of the water vapor.

The relationship between the *actual* absolute humidity at a particular temperature and the *maximum* absolute humidity that can occur at that temperature is called the **relative humidity**. Relative humidity is a ratio between (1) the amount of water vapor in the air, and (2) the amount of water vapor needed to saturate the air at that temperature. The relationship is

$$\frac{\text{actual absolute humidity at present temperature}}{\text{maximum absolute humidity at present temperature}} \times 100\% = \text{relative humidity}.$$

For example, suppose a measurement of the water vapor in the air at 10° C (50° F) finds an absolute humidity of 5.0 g/m³. According to figure 31.1, the maximum amount of water vapor that can be in the air when the temperature is 10° C is about 10 g/m³. The relative humidity is then

$$\frac{5.0 \text{ g/m}^3}{10 \text{ g/m}^3} \times 100\% = 50\%.$$

If the absolute humidity were 10 g/m³, then the air would have all the water vapor it could hold and the relative humidity would be 100%. A relative humidity of 100% means that the air is saturated at the present temperature.

Procedure

Part A: Maximum Amount of Water Vapor

1. Measure the present air temperature in your laboratory room and record it in Data Table 31.1. Use the graph of maximum absolute humidity in figure 31.1 to estimate the *maximum* amount of water vapor a cubic meter of air can hold at this temperature. Since one gram of water has an approximate volume of one milliliter, find the maximum amount of water in liters per cubic meter that can be in the room at the present temperature. Record this maximum, in grams/cubic meter and in liters/cubic meter, in Data Table 31.1.

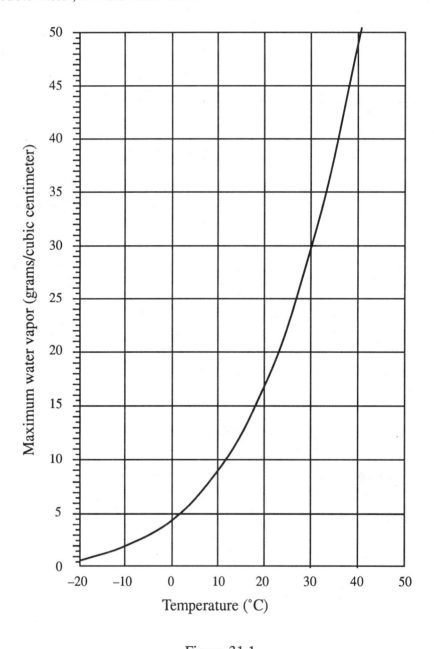

Figure 31.1

2. Measure the length, width, and height of the laboratory room. Record these measurements in Data Table 31.1, then calculate the volume of the room in cubic meters. Record all data and calculations in the data table.

3. Calculate the maximum amount of water vapor the air in the laboratory room can hold at the present temperature. This is found by multiplying the volume of the room (in cubic meters) by the maximum amount of water vapor that could be in the room at the present temperature (in liters per cubic meter). Record all data and calculations in Data Table 31.1.

Part B: Actual Amount of Water Vapor

Evaporation occurs at a rate that is inversely proportional to the relative humidity, ranging from a maximum rate when the air is driest to no net evaporation when the air is saturated. Since evaporation is a cooling process, it is possible to use a thermometer to measure humidity. An instrument called a **psychrometer** has two thermometers, one of which has a damp cloth wick around its bulb end. As air moves past the two thermometer bulbs, the ordinary thermometer (the dry bulb) will measure the present air temperature. Water will evaporate from the wet wick (the wet bulb) until an equilibrium is reached between water vapor leaving the wick and water vapor returning to the wick from the air. Since evaporation lowers the temperature, the depression of the temperature of the wet-bulb thermometer is an indirect measure of the water vapor present in the air. The relative humidity can be determined by obtaining the dry- and wet-bulb temperature readings and referring to a relative humidity chart such as the one found inside the back cover. If the humidity is 100%, no net evaporation will take place from the wet bulb, and both wet- and dry-bulb temperatures will be the same. The lower the humidity, the greater the difference in the temperature reading of the two thermometers.

Relative humidity is a ratio between the actual absolute humidity at a given temperature and the maximum absolute humidity that can occur at that temperature. Knowing the maximum absolute humidity and the relative humidity, you can find the amount of water vapor in the air at the present temperature.

1. Wet the cotton wick on the wet bulb of a sling psychrometer. Whirl the thermometers in the air until the wet-bulb thermometer registers its lowest reading. Record the wet-bulb and dry-bulb temperatures, then use this data to find the relative humidity from the relative humidity chart inside the back cover. Record all data and calculations in the data table.

2. Multiply the humidity (as a fraction) times the maximum amount of water vapor that could be in your laboratory room at the present temperature. Record the amount of water vapor present in the room in the data table.

Results

1. What is the actual amount of water vapor present in the laboratory room air at the present temperature (in g/cm^3)? What is the maximum amount of water vapor that *could* be present?

2. Can the absolute humidity of the air in the room be increased? Explain.

3. Can the relative humidity of the air in the room be increased without adding more water vapor to the air? Explain.

4. Suppose the room air has all the water vapor it will hold at 25° C, and the air is cooled to 15° C. Considering the area of the floor from your measurements, how deep a layer of water will condense from the air?

5. Was the purpose of this lab accomplished? Why or why not? (Your answer to this question should show thoughtful analysis and careful, thorough thinking.)

Data Table 31.1 Amount of Water Vapor in the Laboratory Room	
1. Present temperature of laboratory room air	_____ °C

2. Maximum absolute humidity at present temperature
 	
      In grams per cubic meter
 	
      In liters per cubic meter
	
	 _____ g/m^3
	
	 _____ L/m^3

3. Room length	_____ m
Room width	_____ m
Room height	_____ m
Volume of laboratory room (length × width × height)	_____ m^3
4. Maximum amount of water vapor (row 2 × row 3)	_____ L

5. Dry-bulb reading
 	
      Wet-bulb reading
 	
      Difference in wet- and dry-bulb readings
 	
      Relative Humidity (from relative humidity chart)
	_____ °C
	
	 _____ °C
	
	 _____ °C
	
	 _____ %

6. Amount of water vapor in room (row 5 as decimal × row 4)	_____ L

Name_____Section_____Date_____

Experiment 32: Boyle's Law

Introduction

A gas is not very dense, compared to solids and liquids, and it diffuses rapidly throughout any container in which it is placed. This diffusion is spontaneous and against gravity and only the movement of highly mobile molecules can account for it. The gas molecules must be rapidly moving about, colliding often with other gas molecules and the walls of the container. Since it is not necessary to supply heat to maintain the temperature of a perfectly insulated sample of gas, all of these collisions must be perfectly plastic. On the average, there is no loss of kinetic energy when molecules collide.

The assumptions you can logically make about gases can be extended to explain some of the observable properties of gases. For example, the pressure (force per unit area) that a gas exerts on a surface can be assumed to be the result of the continuous bombardment of gas molecules on the surface. This would explain why adding more air to a tire increases the air pressure in the tire since more air means more bombardment by the added molecules. Likewise, the air pressure increases in your tires when you drive because friction with the road increases the absolute temperature of the air in your tires. Increasing the absolute temperature increases the average molecular kinetic energy, which may increase the frequency of impact on the tire walls as well as the force of impact. In any event, increased frequency and increased force of impact will result in greater air pressure.

In this experiment you will investigate how the volume of a given mass of gas varies with the pressure exerted on it. The pressure and volume of a fixed amount of gas at constant temperature are inversely proportional, which is the relationship known as Boyle's law. In symbols, Boyle's law is

$$P = \frac{k}{V} \quad \text{or} \quad PV = k$$

where P is the gas pressure, V is the volume of the gas, and k is a proportionality constant with a value that depends on the constant temperature and mass of the gas. Suppose V_1 is used to represent the initial volume and V_2 a new volume after a pressure change. Then P_1 and P_2 can be used to represent the initial and new pressures. With the same temperature and mass of gas between the initial and new volume and pressure ($k = k$), then

$$\frac{V_1}{V_2} = \frac{P_1}{P_2} \quad \text{or} \quad P_1 V_1 = P_2 V_2$$

The experimental test of Boyle's law consists of observing a series of volumes, measuring the corresponding pressures, and plotting P versus 1/V to see if a straight line is obtained.

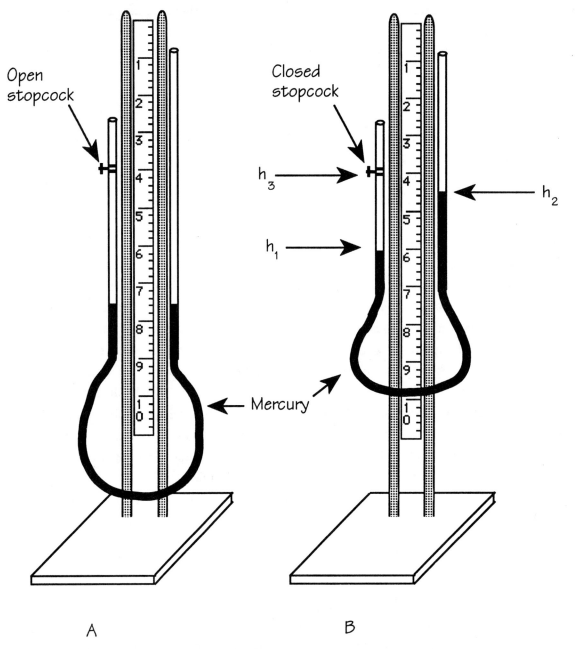

Figure 32.1

Procedure

1. Measure the barometric pressure and record the reading in Data Table 32.1.

2. Inspect the Boyle's law apparatus to make sure that each tube is about half full when the stopcock is open. **CAUTION:** Inform your laboratory instructor of any mercury spills, no matter how small. Any spilled mercury should be cleaned up and properly disposed of before it becomes "lost" in the environment.

3. Adjust the apparatus so the mercury columns are the same height as shown in figure 32.1A, then close the stopcock. The closed side of the apparatus (the side with the stopcock) will remain stationary and should not be touched or handled during the investigation. Note that the mercury columns remain the same height after the stopcock is closed, meaning that atmospheric pressure exists in the closed volume.

4. Measure the top of the closed tube just below the stopcock and record this measurement in Data Table 32.1 as h_3.

5. Move the open side of the tube down as far as you can. This decreases the pressure on the air in the closed tube and the difference in the mercury levels of the two tubes, h_2 and h_1, is the gauge pressure. Record the measurements of h_2 and h_1 in Data Table 32.1, then calculate and record the gauge pressure. Do not move the open tube until after the next procedure step.

6. Since the glass tube has a uniform bore, the volume of air inside the tube is proportional to the length of the tube. The length of the column of air inside the closed tube, $h_3 - h_1$, is therefore representative of the volume of air inside the tube. Calculate the volume of air inside the closed tube and record it in Data Table 32.1.

7. Raise the open tube to the highest position possible in eleven approximately equal steps, recording the h_2 and h_1 for each of the steps. Calculate and record the gauge pressure and the volume of air inside the tube for each of the 12 different observations, recording all calculations in Data Table 32.1.

Results

1. Calculate the total pressure on the confined air in each of the 12 different observations, recording your findings in Data Table 32.1. **Note:** If gauge pressure is indicated in cm, record the barometric pressure also in cm to calculate the total pressure in Data Table 32.1.

2. Calculate PV for each of the 12 different observations, again recording your findings in Data Table 32.1.

3. Calculate 1/V for each of the 12 observations and record your findings in the data table.

4. Plot the total pressure versus the corresponding values of 1/V. Calculate the slope and write it on the graph somewhere and here as well:

5. Describe how the calculations in Data Table 32.1 would verify Boyle's law.

6. Describe how your graph would verify Boyle's law.

7. Does the PV column of Data Table 32.1 agree with your graph? Explain in detail, giving evidence with your answer.

8. Explain how touching or handling the closed tube could introduce error into this experiment. What are other sources of error?

9. Was the purpose of this lab accomplished? Why or why not? (Your answer to this question should show thoughtful analysis and careful, thorough thinking.)

Data Table 32.1   Boyle's Law

Trial	h_1	h_2	Gauge Pressure $(h_2 - h_1)$	Volume $(h_3 - h_1)$	Total Pressure (Atmospheric plus Gauge)	PV	$\frac{1}{V}$
1							
2							
3							
4							
5							
6							

Barometer Reading _____

Measurement at Top of Closed Tube (h_3) _____

Data Table 32.1		Boyle's Law, continued					
Trial	h_1	h_2	Gauge Pressure $(h_2 - h_1)$	Volume $(h_3 - h_1)$	Total Pressure (Atmospheric plus Gauge)	PV	$\frac{1}{V}$
7							
8							
9							
10							
11							
12							

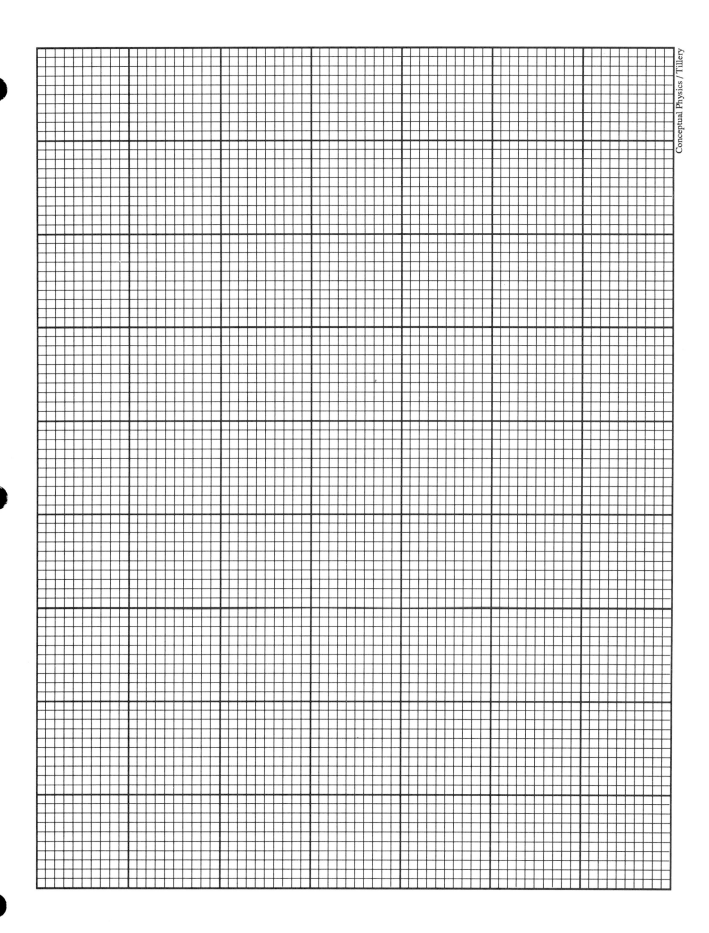

Name_____Section_____Date_____

Experiment 33: Nuclear Radiation

Introduction

All the isotopes of all the elements with an atomic number greater than 83 (bismuth) are radioactive. **Radioactivity** is defined as the *spontaneous emission of particles or energy from an atomic nucleus* as it disintegrates. As a result of the disintegration the nucleus of an atom often undergoes a change of identity, becoming a simpler nucleus. The natural spontaneous disintegration or decomposition of a nucleus is also called **radioactive decay**. Although it is impossible to know when a given nucleus will undergo radioactive decay it is possible to deal with the rate of decay for a given radioactive material with precision.

An unstable nucleus breaks down by radioactive decay to become a more stable nucleus with less energy. There are five common types of radioactive decay and three of these involve alpha, beta, and gamma radiation.

• **Alpha emission.** Alpha (α) emission is the expulsion of an alpha particle from an unstable, disintegrating nucleus. The alpha particle, a helium nucleus, travels from 2 to 12 cm through the air, depending on the energy of emission from the source. An alpha particle is easily stopped by a sheet of paper close to the nucleus. As an example of alpha emission, consider the decay of a radon-222 nucleus, which can be shown by the equation

$$^{222}_{86}Rn \rightarrow {}^{218}_{84}Po + {}^{4}_{2}He.$$

The spent alpha particle eventually acquires two electrons and becomes an ordinary helium atom.

• **Beta emission.** Beta (β⁻) emission is the expulsion of a different particle, a beta particle, from an unstable disintegrating nucleus. A beta particle is simply an electron ejected from the nucleus at a high speed. The emission of a beta particle *increases the number of protons* in a nucleus. It is as if a neutron changed to a proton by emitting an electron, which can be shown as

$$^{1}_{0}n \rightarrow {}^{1}_{1}p + {}^{0}_{-1}e.$$

Carbon-14 is a carbon isotope that decays by beta emission, which can be shown as

$$^{14}_{6}C \rightarrow {}^{14}_{7}N + {}^{0}_{-1}e.$$

Note that the number of protons increased from six to seven, but the mass number remained the same. The mass number is unchanged because the mass of the expelled electron (beta particle) is negligible.

Beta particles are more penetrating than alpha particles and may travel several hundred cm through the air. They can be stopped by a thin layer of metal close to the emitting nucleus, such as a 1-cm thick piece of aluminum. A spent beta particle may eventually join an ion to become part of an atom or it may remain a free electron.

• **Gamma emission**. Gamma (γ) emission is a high-energy burst of electromagnetic radiation from an excited nucleus. It is a burst of light (photon) of wavelength much too short to be detected by the eye. Other types of radioactive decay, such as alpha or beta emission, sometimes leave the nucleus with an excess of energy, a condition called an *excited state*. As in the case of excited electrons, the nucleus returns to a lower energy state by emitting electromagnetic radiation. From a nucleus, this radiation is in the high-energy portion of the electromagnetic spectrum. Gamma is the most penetrating of the three common types of nuclear radiation. Like X rays, gamma rays can pass completely through a person but most gamma radiation can be stopped by a 5-cm thick piece of lead. As other types of electromagnetic radiation, gamma radiation is absorbed by and gives its energy to materials. Since the product nucleus changed from an excited state to a lower energy state, there is no change in the number of nucleons. For example, radon-222 is an isotope that emits gamma radiation:

$$^{222}_{86}Rn^* \rightarrow \,^{222}_{86}Rn + \,^{0}_{0}\gamma$$

(* denotes excited state)

In this investigation you will become acquainted with some of the instrumentation of nuclear physics and examine the behavior of gamma radiation.

Procedure

Part A: Distance and Intensity of Nuclear Radiation

1. Turn on the Geiger counter and adjust it to the voltage range for the tube used. If the tube has a shield, it should be closed to absorb alpha and beta radiation. Thus you will be measuring gamma radiation only in this experiment.

2. Cover the tube with lead foil that has a 3 mm hole and fix the tube in a vertical position as illustrated in figure 33.1. Note: the hole should be centered at the middle of the tube. Read the *background* count in counts per minute (c/m) for 10 minutes and record it in Data Table 33.1. Make sure that all radioactive sources are at the far end of the room while measuring the background.

3. Place a radioactive source on a ring stand directly in line with the hole that is centered on the Geiger tube. Move the source close to the tube to obtain a high count-per-minute reading. Measure and record the distance and a corrected c/m reading in Data Table 33.1. The corrected c/m reading is obtained by subtracting the background count found in procedure step 2.

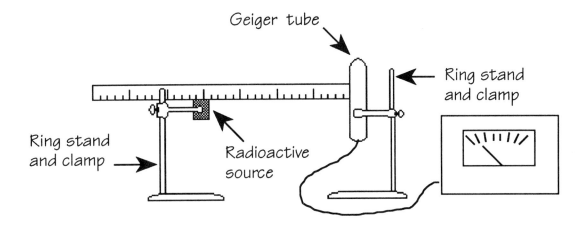

Figure 33.1

4. Move the source away from the Geiger counter tube in 1 cm intervals, recording the corrected c/m reading and the distance each time. Continue this process until there is no change for three successive intervals.

5. Plot the corrected c/m readings versus distance on a full sheet of graph paper.

Part B: Absorption of Nuclear Radiation

1. Remove the lead foil from the Geiger tube, but leave the tube mounted in a vertical position. The tube measures alpha, beta, and gamma with the metal window open. It measures gamma only with the metal window closed. Adjust the voltage of the Geiger tube, then measure the background with the metal window closed. Record the background count in Data Table 33.2.

2. Place a radioactive source close enough to the tube for a high c/m reading. Record the reading, corrected for background, in Data Table 33.2.

3. Place a single sheet of lead foil between the radioactive source and the tube. The lead sheet should be as close to the radioactive source as possible. Record the corrected c/m reading for the attenuation property of a single sheet of lead.

4. Add additional sheets of lead foil close to the radioactive source, recording the corrected readings in Data Table 33.2.

5. Use a full sheet of graph paper to plot corrected c/m readings versus number of lead sheets.

6. Your instructor might have other materials that can be studied concerning the passage of alpha, beta, and gamma rays.

Results

1. Describe how the behavior in count rate changes as a function of distance from a radioactive gamma source.

2. Generalize your description of c/m and distance by stating a quantitative relationship describing how the intensity of gamma radiation varies with distance from the source.

3. Would the generalization of question 2 apply to alpha and beta radiation as well as gamma? Explain why or why not.

4. What is the behavior in count rate as a function of thickness of a material for gamma rays?

5. How could you compensate if the actual count rate changes during a long counting period?

6. How does the penetrating ability of beta rays compare with that of gamma rays? How about alpha particles? Be quantitative in your answer.

7. Was the purpose of this lab accomplished? Why or why not? (Your answer to this question should show thoughtful analysis and careful, thorough thinking.)

Data Table 33.1   Distance and Intensity of Nuclear Radiation

Background Count _____ c/m

Distance (m)	Intensity (c/m)	Corrected Intensity (c/m)
_____	_____	_____
_____	_____	_____
_____	_____	_____
_____	_____	_____
_____	_____	_____
_____	_____	_____
_____	_____	_____
_____	_____	_____
_____	_____	_____

Data Table 33.2   Distance and Intensity of Nuclear Radiation

Background Count _____ c/m

Number of Sheets	Intensity (c/m)	Corrected Intensity (c/m)
_____	_____	_____
_____	_____	_____
_____	_____	_____
_____	_____	_____
_____	_____	_____
_____	_____	_____
_____	_____	_____
_____	_____	_____
_____	_____	_____

Appendix I: The Simple Line Graph

An equation describes a relationship between variables, and a graph helps you "picture" this relationship. A line graph pictures how changes in one variable go with changes in a second variable; that is, how the two variables change together. One variable usually can be easily manipulated; the other variable is caused to change in value by manipulation of the first variable. The *manipulated* variable is known by various names (*independent, input, or cause variable*) and the *responding* variable is known by various related names (*dependent, output, or effect variable*). The manipulated variable is usually placed on the horizontal or *x*-axis of the graph, so you can also identify it as the *x-variable*. The responding variable is placed on the vertical or *y*-axis. This variable is identified as the *y-variable*.

The graph in appendix figure I.1 shows the mass of different volumes of water at room temperature. Volume is placed on the *x*-axis because the volume of water is easily manipulated and the mass values change as a consequence of changing the values of volume. Note that both variables are named, and the measuring unit for each variable is identified on the graph.

The graph also shows a number scale on each axis that represents changes in the values of each variable. The scales are usually, but not always, linear. A *linear* scale has equal intervals that represent equal increases in the value of the variable. Thus a certain distance on the *x*-axis to the right represents a certain increase in the value of the *x*-variable. Likewise, certain distances up the *y*-axis represent certain increases in the value of the *y*-variable. In the example, each mark has a value of

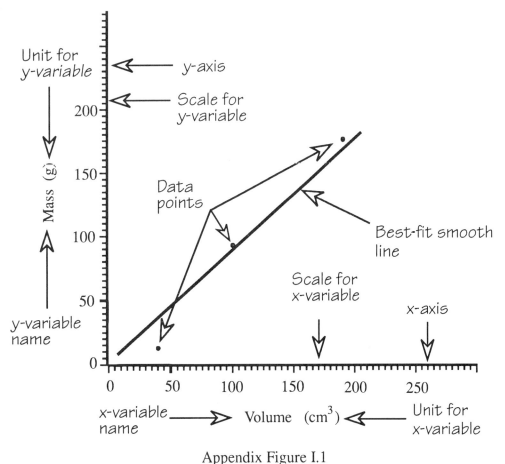

Appendix Figure I.1

five. Scales are usually chosen in such a way that the graph is large and easy to read. The *origin* is the only point where both the *x*- and *y*-variables have a value of zero at the same time.

The example graph has three data points. A *data point* represents measurements of two related variables that were made at the same time. For example, a volume of 190 cm^3 of water was found to have a mass of 175 g. Locate 190 cm^3 on the *x*-axis and imagine a line moving straight up from this point on the scale (each mark on the scale has a value of 5 cm^3). Now locate 175 g on the *y*-axis and imagine a line moving straight out from this point on the scale (again, note that each mark on this scale has a value of 5 g). Where the lines meet is the data point for the 190 cm^3 and 175 g measurements. A data point is usually indicated with a small dot or an x; a dot is used in the example graph.

A "best-fit" smooth, straight line is drawn as close to all the data points as possible. If it is not possible to draw the straight line *through* all the data points (and it usually never is), then a straight line should be drawn that has the same number of data points on both sides of the line. Such a line will represent a best approximation of the relationship between the two variables. The *origin* is also used as a data point in the example because a volume of zero will have a mass of zero. In any case, the dots are *never* connected as in dot-to-dot sketches. For most of the experiments in this lab manual a set of perfect, error-free data would produce a straight line. In such experiments it is not a straight line because of experimental error, and you are trying to eliminate the error by approximating what the relationship should be.

The smooth, straight line tells you how the two variables get larger together. If the scales on both the axes are the same, a 45° line means that the two variables are increasing in an exact direct proportion. A more flat or more upright line means that one variable is increasing faster than the other. The more you work with graphs, the easier it will become for you to analyze what the slope means.

Appendix II: The Slope of a Straight Line

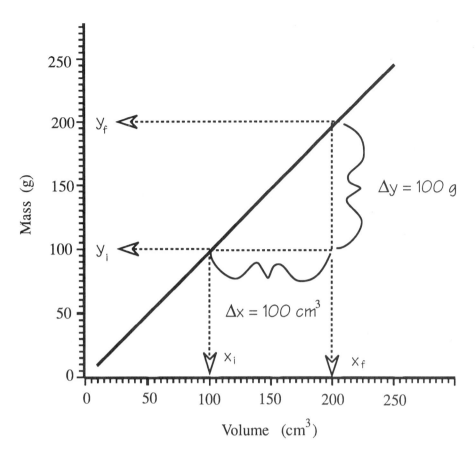

Appendix Figure II.1

One way to determine the relationship between two variables that are graphed with a straight line is to calculate the **slope.** The slope is a ratio between the changes in one variable and the changes in the other. The ratio is between the changes in the value of the *x*-variable compared to the changes in the value of the *y*-variable. The symbol Δ (Greek letter delta) means "change in," so the symbol Δx means "change in *x*." The first step in calculating the slope is to find out how much the *x*-variable is changing (Δx) in relation to how much the *y*-variable is changing (Δy). You can find this relationship by first drawing a dashed line to the right of the straight line so that the *x*-variable has increased by some convenient unit as shown in the example in appendix figure II.1. *Where you start or end this dashed line will not matter since the ratio between the variables will be the same everywhere on the graph line.* However, it is very important to remember when finding a slope of a graph to *avoid using data points* in your calculations. Two points whose coordinates are easy to find should be used instead of data points. One of the main reasons for plotting a graph and drawing a best-fit straight line is to smooth out any measurement errors made. Using data points directly in calculations defeats this purpose.

The Δx is determined by subtracting the final value of the *x*-variable on the dashed line (x_f) from the initial value of the *x*-variable on the dashed line (x_i), or $\Delta x = x_f - x_i$. In the example graph above, the dashed line has an x_f of 200 cm³ and an x_i of 100 cm³, so Δx is 200 cm³ − 100 cm³, or 100 cm³. *Note that Δx has both a number value and a unit.*

Now you need to find Δy. The example graph shows a dashed line drawn back up to the graph line from the x-variable dashed line. The value of Δy is $y_f - y_i$. In the example, $\Delta y = 200$ g $- 100$ g. The slope of a straight graph line is the ratio of Δy to Δx, or

$$\text{Slope} = \frac{\Delta y}{\Delta x}.$$

In the example,

$$\text{Slope} = \frac{100 \text{ g}}{100 \text{ cm}^3}$$

or

$$\text{Slope} = 1 \text{ g} / \text{cm}^3.$$

Thus the slope is 1 g/cm³ and this tells you how the variables change together. Since g/cm³ is also the definition of density, you have just calculated the density of water from a graph.

Note that the slope can be calculated only for two variables that are increasing together (variables that are in direct proportion and have a line that moves upward and to the right). If variables change in any other way, mathematical operations must be performed to *change the variables into this relationship*. Examples of such necessary changes include taking the inverse of one variable, squaring one variable, taking the inverse square, and so forth.

Appendix III: Experimental Error

All measurements are subject to some uncertainty, as a wide range of errors can and do happen. Measurements should be made with great accuracy and with careful thought about what you are doing to reduce the possibility of error. Here is a list of some of the possible sources of error to consider and avoid.

Improper Measurement Technique. Always use the smallest division or marking on the scale of the measuring instrument, then estimate the next interval between the shown markings. For example, the instrument illustrated in appendix figure III.1 shows a measurement of 2.45 units, and the .05 is estimated because the reading is about halfway between the marked divisions of 2.4 and 2.5. If you do not estimate the next smallest division you are losing information that may be important to the experiment you are conducting.

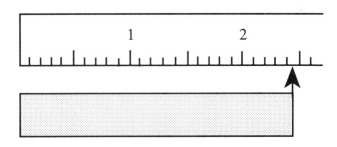

Appendix Figure III.1

Incorrect Reading. This is an error in reading (misreading) an instrument scale. Some graduated cylinders, for example, are calibrated with marks that represent 2.0 mL intervals. Believing that the marks represent 1.0 mL intervals will result in an incorrect reading. This category of errors also includes the misreading of a scale that often occurs when you are not paying sufficient attention to what you are doing.

Incorrect Recording. A personal mistake that occurs when the data are incorrectly recorded; for example, making a reading of 2.54 units and then recording a measurement of 2.45 units.

Assumptions About Variables. A personal mistake that occurs when there is a lack of clear, careful thinking about what you are doing. Examples are an assumption that water always boils at a temperature of 212° F (100° C), or assuming that the temperature of a container of tap water is the same now as it was 15 minutes ago.

Not Controlling Variables. This category of errors is closely related to the assumptions category but in this case means failing to recognize the influence of some variable on the outcome of an experiment. An example is the failure to recognize the role that air resistance might have in influencing the length of time that an object falls through the air.

Math Errors. This is a personal error that happens to everyone but penalizes only those who do not check their work and think about the results and what they mean. Math errors include not using significant figures for measurement calculations.

Accidental Blunders. Just like math errors, accidents do happen. However, the blunder can come from a poor attitude or frame of mind about the quality of work being done. In the laboratory, an example of a lack of quality work would be spilling a few drops of water during an experiment with an "Oh well, it doesn't matter" response.

Instrument Calibration. Errors can result from an incorrectly calibrated instrument, but these errors can be avoided by a quality work habit of checking the calibration of an instrument against a known standard, then adjusting the instrument as necessary.

Inconsistency. Errors from inconsistency are again closely associated with a lack of quality work habits. Such errors could result from a personal bias; that is, trying to "fit" the data to an expected outcome, or using a single measurement when a spread of values is possible.

Whatever the source of errors, it is important that you recognize the error, or errors, in an experiment and know the possible consequence and impact on the results. After all, how else will you know if two seemingly different values from the same experiment are acceptable as the "same" answer or which answer is correct? One way to express the impact of errors is to compare the results obtained from an experiment with the true or accepted value. Everyone knows that percent is a ratio that is calculated from

$$\frac{\text{Part}}{\text{Whole}} \times 100\% \text{ of whole} = \% \text{ of part}.$$

This percent relationship is the basic form used to calculate a percent error or a percent difference.

The **percent error** is calculated from the *absolute difference* between the experimental value and the accepted value (the part) divided by the accepted value (the whole). Absolute difference is designated by the use of two vertical lines around the difference, so

$$\% \text{ Error} = \frac{|\text{Experimental value} - \text{Accepted value}|}{\text{Accepted value}} \times 100\%.$$

Note that the absolute value for the part is obtained when the smaller value is subtracted from the larger. For example, suppose you experimentally determine the frequency of a tuning fork to be 511 Hz but the accepted value stamped on the fork is 522 Hz. Subtracting the smaller value from the larger, the percentage error is

$$\frac{|522 \text{ Hz} - 511 \text{ Hz}|}{522 \text{ Hz}} \times 100\% = 2.1\%.$$

You should strive for the lowest percentage error possible, but some experiments will have a higher level of percentage errors than other experiments, depending on the nature of the measurements required. In some experiments the acceptable percentage error might be 5%, but other experiments could require a percentage error of no more than 2%.

A true, or accepted, value is not always known so it is sometimes impossible to calculate an actual error. However, it is possible in these situations to express the error in a measured quantity as a percent of the quantity itself. This is called a **percent difference**, or a percent deviation from the mean. This method is used to compare the accuracy of two or more measurements by seeing how consistent they are with each other. The percent difference is calculated from the *absolute difference* between one measurement and a second measurement, divided by the average of the two measurements. As before, absolute difference is designated by the use of two vertical lines around the difference, and

$$\% \text{ Difference} = \frac{|\text{One value} - \text{Another value}|}{\text{Average of the two values}} \times 100\%.$$

Appendix IV: Significant Figures

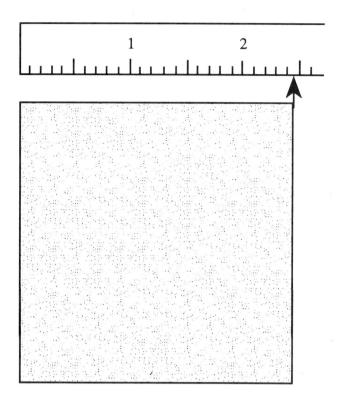

The numerical value of any measurement will always contain some uncertainty. Suppose, for example, that you are measuring one side of a square piece of paper as shown above. You could say that the paper is *about* 2.5 cm wide and you would be correct. This measurement, however, would be unsatisfactory for many purposes. It does not approach the true value of the length and contains too much uncertainty. It seems clear that the paper width is larger than 2.4 cm but shorter than 2.5 cm. But how much larger than 2.4 cm? You cannot be certain if the paper is 2.44, 2.45, or 2.46 cm wide. As your best estimate, you might say that the paper is 2.45 cm wide. Everyone would agree that you can be certain about the first two numbers (2.4) and they should be recorded. The last number (0.05) has been estimated and is not certain. The two certain numbers, together with one uncertain number, represent the greatest accuracy possible with the ruler being used. The paper is said to be 2.45 cm wide.

A **significant figure** is a number that is believed to be correct with some uncertainty only in the last digit. The value of the width of the paper (2.45 cm) represents three significant figures. As you can see, the number of significant figures can be determined by the degree of accuracy of the measuring instrument being used. But suppose you need to calculate the area of the paper. You would multiply 2.45 cm × 2.45 cm and the product for the area would be 15.24635 cm^2. This is a greater accuracy than you were able to obtain with your measuring instrument. *The result of a calculation can be no more accurate than the values being treated*. Because the measurement had only three significant figures (two certain, one uncertain), then the answer can have only three significant figures. Thus the area is correctly expressed as 15.2 cm^2.

There are a few simple rules that will help you determine how many significant figures are contained in a reported measurement.

Rule 1. All digits reported as a direct result of a measurement are significant.

Rule 2. Zero is significant when it occurs between nonzero digits. For example, 607 has three significant figures and the zero is one of the significant figures.

Rule 3. In figures reported as *larger than the digit one,* the digit zero is *not significant* when it follows a nonzero digit to indicate place. For example, in a report that "23,000 people attended the rock concert," the digits 2 and 3 are significant but the zeros are not significant. In this situation the 23 is the measured part of the figure and the three zeros tell you an estimate of how many attended the concert, that is, 23 thousand. If the figure is a measurement rather than an estimate, then it is written *with a decimal point after the last zero* to indicate that the zeros *are* significant. Thus 23,000 has *two* significant figures (2 and 3), but 23,000. has *five* significant figures. The figure 23,000 means "about 23 thousand" but 23,000. means 23,000. and not 22,999 or 23,001. One way to show the number of significant figures is to use scientific notation, e.g., 2.3×10^3 has two significant figures, 2.30×10^3 has three, and 2.300×10^4 has four significant figures. Another way to show the number of significant figures is to put a bar over the top of a significant zero if it could be mistaken for a place-holder.

Rule 4. In figures reported as *smaller than the digit one*, zeros after a decimal point that come before nonzero digits *are not* significant and serve only as place-holders. For example, 0.0023 has two significant figures, 2 and 3. Zeros alone after a decimal point or zeros after a nonzero digit indicate a measurement, however, so these zeros *are* significant. The figure 0.00230, for example, has three significant figures since the 230 means 230 and not 229 or 231. Likewise, the figure 3.000 cm has four significant figures because the presence of the three zeros means that the measurement was actually 3.000 and not 2.999 or 3.001.

Multiplication and Division

When multiplying or dividing measurement figures, the answer may have no more significant figures than the *least* number of significant figures in the figures being multiplied or divided. This simply means that an answer can be no more accurate than the least accurate measurement entering into the calculation, and that you cannot improve the accuracy of a measurement by doing a calculation. For example, in multiplying 54.2 mi/hr × 4.0 hours to find out the total distance traveled, the first figure (54.2) has three significant figures but the second (4.0) has only two significant figures. The answer can contain only two significant figures since this is the weakest number of those involved in the calculation. The correct answer is therefore 220 miles, not 216.8 miles. This may seem strange since multiplying the two numbers together gives the answer of 216.8 miles. This answer, however, means a greater accuracy than is possible and the accuracy cannot be improved over the weakest number involved in the calculation. Since the weakest number (4.0) has only two significant figures the answer must also have only two significant figures, which is 220 miles.

The result of a calculation is **rounded** to have the same least number of significant figures as the least number of a measurement involved in the calculation. When rounding numbers, the last significant figure is increased by one if the number after it is five or larger. If the number after the last significant figure is four or less, the nonsignificant figures are simply dropped. Thus, if two significant figures are called for in the answer of the above example, 216.8 is rounded up to 220 because the last number after the two significant figures is 6, a number larger than 5. If the calculation result had been 214.8, the rounded number would be 210 miles.

Note that *measurement figures* are the only figures involved in the number of significant figures in the answer. Numbers that are **counted or defined** are not included in the determination of significant figures in an answer. For example, when dividing by 2 to find an average property of two objects, the 2 is ignored when considering the number of significant figures. Defined numbers are defined exactly and are not used in significant figures. For example, that a diameter is 2 times the radius is not a measurement. In addition, 1 kilogram is *defined* to be exactly 1000 grams so such a conversion is not a measurement.

Addition and Subtraction

Addition and subtraction operations involving measurements, as with multiplication and division, cannot result in an answer that implies greater accuracy than the measurements had before the calculation. Recall that the last digit in a measurement is considered to be uncertain because it is the result of an estimate. The answer to an addition or subtraction calculation can have this uncertain number *no farther from the decimal place than it was in the weakest number involved in the calculation*. Thus when 8.4 is added to 4.926, the weakest number is 8.4 and the uncertain number is .4, one place to the right of the decimal. The sum of 13.326 is therefore rounded to 13.3, reflecting the placement of this weakest doubtful figure.

Example Problem

In appendix III, "Experimental Error," an example was given of an experimental result of 511 Hz and an accepted value of 522 Hz, resulting in a calculation of

$$\frac{|522 \text{ Hz} - 511 \text{ Hz}|}{522 \text{ Hz}} \times 100\% = 2.1\%.$$

Since 522 − 511 is 11, the least number of significant figures of measurements involved in this calculation is *two*. Note that the 100 does not enter into the determination since it is not a measurement number. The calculated result (from a calculator) is 2.1072797, which is rounded off to have only two significant figures, so the answer is recorded as 2.1%.

Appendix V: Conversion of Units

The measurement of most properties results in both a numerical value and a unit. The statement that a glass contains 50 cm^3 of a liquid conveys two important concepts — the numerical value of 50, and the reference unit of cubic centimeters. Both the numerical value and the unit are necessary to communicate correctly the volume of the liquid.

When working with calculations involving measurement units, *both* the numerical value and the units are treated mathematically. As in other mathematical operations, there are general rules to follow.

Rule 1. Only properties with *like units* may be added or subtracted. It should be obvious that adding quantities such as 5 dollars and 10 dimes is meaningless. You must first convert to like units before adding or subtracting.

Rule 2. Like or unlike units may be multiplied or divided and treated in the same manner as numbers. You have used this rule when dealing with area (length $\times$ length = length^2 or cm $\times$ cm = cm^2) and when dealing with volume (length $\times$ length $\times$ length = length^3 or cm $\times$ cm $\times$ cm = cm^3).

You can use the above two rules to create a **conversion ratio** that will help you change one unit to another. Suppose you need to convert 2.3 kilograms to grams. First, write the relationship between kilograms and grams:

$$1000 \text{ grams} = 1.000 \text{ kg}.$$

Next, divide both sides by what you wish to convert *from* (kilograms in this example):

$$\frac{1000 \text{ g}}{1.000 \text{ kg}} = \frac{1.000 \text{ kg}}{1.000 \text{ kg}}.$$

One kilogram divided by one kilogram equals 1, just as 10 divided by 10 equals 1. Therefore, the right side of the relationship becomes 1:

$$\frac{1000 \text{ g}}{1.000 \text{ kg}} = 1.$$

The 1 is usually understood — that is, not stated — and the operation is called *canceling*. Canceling leaves you with the fraction 1000 g/1.000 kg, which is a conversion ratio that can be used to convert from kg to g. You simply multiply the conversion ratio by the numerical value and unit you wish to convert:

$$2.3 \text{ kg} \times \frac{1000 \text{ g}}{1.000 \text{ kg}} = 2300 \text{ g}.$$

The kg units cancel. Showing the whole operation with units only, you can see how you end up with the correct unit of g:

$$\text{kg} \times \frac{\text{g}}{\text{kg}} = \frac{\text{kg} \cdot \text{g}}{\text{kg}} = \text{g}.$$

Since you did obtain the correct unit, you know that you used the correct conversion ratio. If you had blundered and used an inverted conversion ratio, you would obtain:

$$2.3 \times \frac{1.000 \text{ kg}}{1000 \text{ g}} = 23 \frac{\text{kg}^2}{\text{g}},$$

which yields the meaningless, incorrect units of kg^2/g. Carrying out the mathematical operations on the numbers and the units will always tell you if you used the correct conversion ratio or not.

Example Problem

A distance is reported as 100.0 km and you want to know how far this is in miles.

Solution

First, you need to obtain a conversion factor from a textbook or reference book, which usually groups similar conversion factors in a table. Such a table will show two conversion factors for kilometers and miles: (a) 1.000 km = 0.621 mi, and (b) 1.000 mi = 1.609 km. You select the factor that is in the same form as your problem. For example, your problem is 100.0 km = ? mi. The conversion factor in this form is 1.000 km = 0.621 mi.

Second, you convert this conversion factor into a **conversion ratio** by dividing the factor by what you want to convert *from*.

Conversion factor: 1.000 km = 0.621 mi

Divide factor by what you want to convert from: $\dfrac{1.000 \text{ km}}{1.000 \text{ km}} = \dfrac{0.621 \text{ mi}}{1.000 \text{ km}}$

Resulting conversion ratio: $\dfrac{0.621 \text{ mi}}{\text{km}}$

The conversion ratio is now multiplied by the numerical value and unit you wish to convert.

$$100.0 \text{ km} \times \frac{0.621 \text{ mi}}{\text{km}}$$

$$100.0 \times 0.621 \; \frac{\text{km} \cdot \text{mi}}{\text{km}}$$

62.1 miles.

Appendix VI: Scientific Notation

Most of the properties of things that you might measure in your everyday world can be expressed with a small range of numerical values together with some standard unit of measure. The range of numerical values for most everyday things can be dealt with by using units (1's), tens (10's), hundreds (100's), or perhaps thousands (1,000's). But the universe contains some objects of incredibly large size that require some very big numbers to describe. The sun, for example, has a mass of about 1,970,000,000,000,000,000,000,000,000,000 kg. On the other hand, very small numbers are needed to measure the size and parts of an atom. The radius of a hydrogen atom, for example, is about 0.00000000005 m. Such extremely large and small numbers are cumbersome and awkward since there are so many zeros to keep track of, even if you are successful in carefully counting all the zeros. A method does exist to deal with extremely large or small numbers in a more condensed form. The method is called **scientific notation**, but it is also sometimes called *powers of ten* or *exponential notation* since it is based on exponents of 10. Whatever it is called, the method is a compact way of dealing with numbers that not only helps you keep track of zeros but also provides a simplified way to make calculations as well.

In algebra you save a lot of time (as well as paper) by writing ($a \times a \times a \times a \times a$) as a^5. The small number written to the right and above a letter or number is a superscript called an **exponent**. The exponent means that the letter or number is to be multiplied by itself that many times. For example, a^5 means "a" multiplied by itself five times, or $a \times a \times a \times a \times a$. As you can see, it is much easier to write the exponential form of this operation than it is to write out the long form.

Scientific notation uses an exponent to indicate the power of the base 10. The exponent tells how many times the base, 10, is multiplied by itself. For example:

$$10,000. = 10^4$$

$$1,000. = 10^3$$

$$100. = 10^2$$

$$10. = 10^1$$

$$1. = 10^0$$

$$0.1 = 10^{-1}$$

$$0.01 = 10^{-2}$$

$$0.001 = 10^{-3}$$

$$0.0001 = 10^{-4}$$

This table could be extended indefinitely, but this somewhat shorter version will give you an idea of how the method works. The symbol 10^4 is read as "ten to the fourth power" and means $10 \times 10 \times 10 \times 10$. Ten times itself four times is 10,000, so 10^4 is the scientific notation for 10,000. It is also equal to the number of zeros between the 1 and the decimal point. That is, to write the longer form of 10^4 you simply write 1, then move the decimal point four places to the *right*; hence ten to the fourth power is 10,000.

The powers of ten table also shows that numbers smaller than one have negative exponents. A negative exponent means a reciprocal:

$$10^{-1} = \frac{1}{10} = 0.1$$

$$10^{-2} = \frac{1}{100} = 0.01$$

$$10^{-3} = \frac{1}{1000} = 0.001$$

To write the longer form of 10^{-4}, you simply write 1 then move the decimal point four places to the *left;* hence ten to the negative four power is 0.0001.

Scientific notation usually, but not always, is expressed as the product of two numbers: (1) a number between 1 and 10 that is called the **coefficient**, and (2) a power of ten that is called the **exponential**. For example, the mass of the sun that was given in long form earlier is expressed in scientific notation as

$$1.97 \times 10^{30} \text{ kg}$$

and the radius of a hydrogen atom is

$$5.0 \times 10^{-11} \text{ m.}$$

In these expressions, the coefficients are 1.97 and 5.0 and the power of ten notations are the exponentials. Note that in both of these examples, the exponential tells you where to place the decimal point if you wish to write the number all the way out in the long form. Sometimes scientific notation is written without a coefficient, showing only the exponential. In these cases the coefficient of 1.0 is understood; that is, not stated. If you try to enter a scientific notation in your calculator, however, you will need to enter the understood 1.0 or the calculator will not be able to function correctly. Note also that 1.97×10^{30} kg and the expressions 0.197×10^{31} kg and 19.7×10^{29} kg are all correct expressions of the mass of the sun. By convention, however, you will use the form that has one digit to the left of the decimal.

Example Problem

What is 26,000,000 in scientific notation?

Solution

Count how many times you must shift the decimal point until one digit remains to the left of the decimal point. For numbers larger than the digit 1, the number of shifts tells you how much the exponent is increased, so the answer is 2.6×10^7, which means the coefficient 2.6 is multiplied by 10 seven times.

Example Problem

What is 0.000732 in scientific notation? (Answer: 7.32×10^{-4}.)

Multiplication and Division

It was stated earlier that scientific notation provides a compact way of dealing with very large or very small numbers but provides a simplified way to make calculations as well. There are a few mathematical rules that will describe how the use of scientific notation simplifies these calculations.

To *multiply* two scientific notation numbers, the coefficients are multiplied as usual and the exponents are *added* algebraically. For example, to multiply (2×10^2) by (3×10^3), first separate the coefficients from the exponentials,

$$(2 \times 3) \times (10^2 \times 10^3),$$

then multiply the coefficients and add the exponents,

$$6 \times 10^{(2+3)} = 6 \times 10^5.$$

Adding the exponents is possible because $10^2 \times 10^3$ means the same thing as $(10 \times 10) \times (10 \times 10 \times 10)$, which equals $(100) \times (1,000)$, or 100,000, which is expressed as 10^5 in scientific notation. Note that two negative exponents add algebraically, for example, $10^{-2} \times 10^{-3} = 10^{[(-2)+(-3)]} = 10^{-5}$. A negative and a positive exponent also add algebraically, as in $10^5 \times 10^{-3} = 10^{[(+5)+(-3)]} = 10^2$.

If the result of a calculation involving two scientific notation numbers does not have the conventional one digit to the left of the decimal, move the decimal point so it does, changing the exponent according to which way and how much the decimal point is moved. Note that the exponent increases by one number for each decimal point moved to the left. Likewise, the exponent decreases by one number for each decimal point moved to the right. For example, $938. \times 10^3$ becomes 9.38×10^5 when the decimal point is moved two places to the left.

To *divide* two scientific notation numbers, the coefficients are divided as usual and the exponents are *subtracted*. For example, to divide (6×10^6) by (3×10^2), first separate the coefficients from the exponentials,

$$\left(\frac{6}{3}\right) \times \left(\frac{10^6}{10^2}\right)$$

then divide the coefficients and subtract the exponents,

$$2 \times 10^{(6-2)} = 2 \times 10^4$$

Note that when you subtract a negative exponent, for example, $10^{[(3)-(-2)]}$, you change the sign and add, $10^{(3+2)} = 10^5$.

Example Problem

Solve the following problem concerning scientific notation:

$$\frac{(2 \times 10^4) \times (8 \times 10^{-6})}{8 \times 10^4}.$$

Solution

First, separate the coefficients from the exponentials,

$$\frac{2 \times 8}{8} \times \frac{10^4 \times 10^{-6}}{10^4},$$

then multiply and divide the coefficients and add and subtract the exponents as the problem requires,

$$2 \times 10^{\{[(4)+(-6)]-[4]\}}.$$

Solving these remaining operations gives 2×10^{-6}.